KT-558-249

Environmental Politics

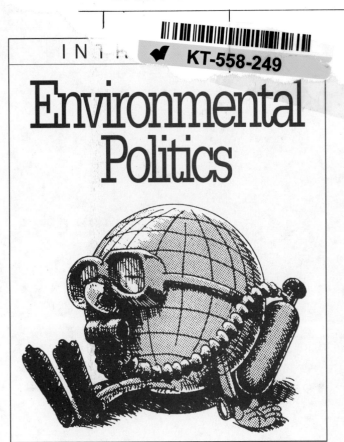

Stephen Croall and William Rankin

Edited by Richard Appignanesi

ICON ... TOTEM BOOKS USA

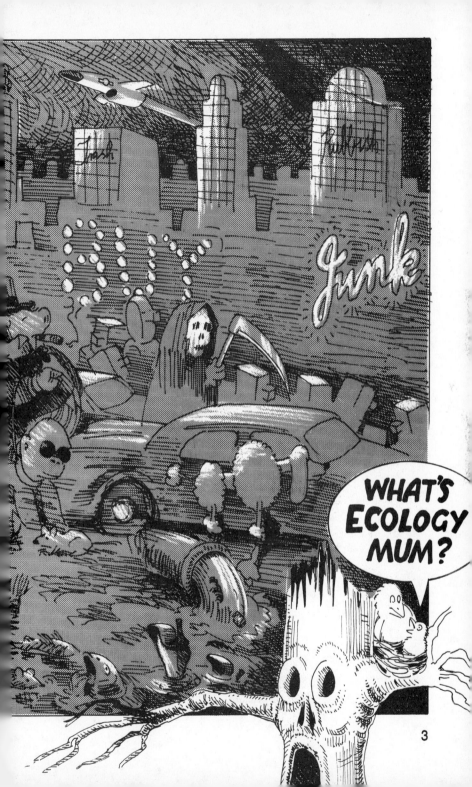

3

This edition published in the UK
in 2000 by Icon Books Ltd.,
Grange Road, Duxford,
Cambridge CB2 4QF
email: info@iconbooks.co.uk
www.iconbooks.co.uk

Distributed in the UK, Europe,
Canada, South Africa and Asia by the
Penguin Group: Penguin Books Ltd.,
27 Wrights Lane, London W8 5TZ

This edition published in Australia
in 2000 by Allen & Unwin Pty. Ltd.,
PO Box 8500, 9 Atchison Street,
St. Leonards NSW 2065

Previously published in the UK and
Australia in 1992 under the title
Ecology for Beginners

Reprinted 1993, 1994, 1995, 1997

Published in the United States
in 2000 by Totem Books
Inquiries to: PO Box 223,
Canal Street Station,
New York, NY 10013

In the United States,
distributed to the trade by
National Book Network Inc.,
4720 Boston Way, Lanham,
Maryland 20706

Library of Congress catalog
card number applied for

ISBN 1 84046 159 4

Originating editor: Richard Appignanesi

Printed and bound in Australia
by McPherson's Printing Group, Victoria

*'The writing of books
is the epidemical
conspiracy for the destruction of paper.'*
- Dr. Johnson

Environmental Politics – the struggle to negotiate between competing interests in determining environmental conditions of existence; the relation of living organisms to their native habitat; *see also* **conservation**, **ecology**, **environmentalism** …

The story so far . . .

Planet Earth condensed 4,600 million years ago from hot gases and cosmic dust. It cooled into a beautiful blue orb, slightly squashed at one pole but still easy to fall in love with. Barring accidents, like the sun going out, it will be here for another 10,000 million years.

It may seem big to us but Earth is a tiny speck in a universe so huge that your brain hurts just thinking about it.

But where on earth did the universe come from?

Life

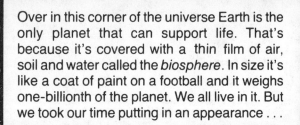

Over in this corner of the universe Earth is the only planet that can support life. That's because it's covered with a thin film of air, soil and water called the *biosphere*. In size it's like a coat of paint on a football and it weighs one-billionth of the planet. We all live in it. But we took our time putting in an appearance . . .

An commercial Artist's Impression of the Arrival of the First Organisms

The first organisms lived on sulphur . . .

Then some of them began giving off oxygen and the air-breathers inherited the Earth . . .

These bacteria gave off 'waste' that in time became fossil and mineral deposits. Some bacteria evolved into plants while others chose to become animals.

9

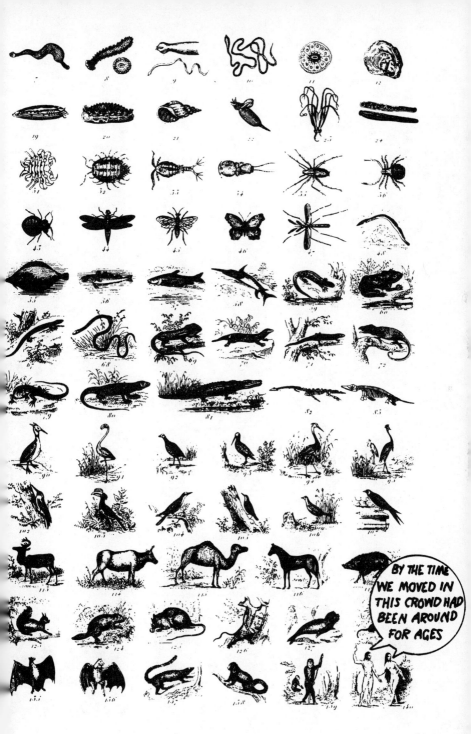

Imagining the history of the Earth as one month,* human beings in their present form (*Homo sapiens*) have only been here ...

... FOR A MINUTE OR SO

BETTER LATE THAN NEVER!

SOMEWHERE IN AFRICA

OR POSSIBLY EVEN CHINA

THAT'S A MATTER OF OPINION

PERHAPS WE'RE NATURE'S FIRST MISTAKE

✳ 1 day = 150 million years

Personally, we're inclined to agree with Mao Tse-tung...

OF ALL THINGS PEOPLE ARE THE MOST PRECIOUS

FRANKLY I PREFER OWLS, AND THE OCCASIONAL MOUSE

Early Man and Woman behaved themselves quite well.

The world had only five or ten million inhabitants in those days so there was plenty of elbow-room. They hunted, slashed-and-burned, fished, gathered, and ate each other for 99% of human history.

The Neolithic Revolution

SEVEN seconds ago on our historical calendar a lot of people tired of chasing about being good ecologists. They started putting down roots.

14

The invention of agriculture was a turning point in human history. The hunters and nomadic herdsmen became 'marginal' and only survived in places that no-one else wanted: the African bush, the Australian outback, the arctic regions of America and Siberia, etc.

They had left nature largely undisturbed. The peasants who succeeded them shaped the environment, integrating themselves with nature. They intervened directly in the *ecosystem*.

The Ecosystem

The ecosystem is the complex web linking animals, plants and other life forms in any particular environment, whether it's your windowbox or the whole biosphere. Everything hangs together in the ecosystem — alter one part and you alter the others. Sooner or later.

Humans are only one factor in the ecosystem. But we don't quite see it that way. We set ourselves apart and call all the other factors or species *natural resources* – or simply *nature*.

Human survival depends upon preserving the ecosystem. It's the boundary of existence, the framework of human activity. The ecosystem can do without us but we can't do without the ecosystem. As long as we live on Earth.

Viewing nature as no more than a bundle of resources for human consumption is inviting trouble.

Commoner's Laws of Ecology:
1. Everything is connected with everything else.
2. Everything must go somewhere.
3. Nature knows best.
4. There's no such thing as a free lunch.

Civilisation

Agriculture spread from southern
Asia and the Middle East
throughout the Eurasian and
African continents, and through
the Americas from Mexico and
northern South America.

People swapped the wandering
life for a roof over their heads,
tilled the soil and kept animals.
Peasants traded surpluses.
Temporary markets became
permanent settlements . . . which
grew . . . and grew . . . and the real
problems began.

and now follows
A Disastrous
History of
Ecology

IN A
SOCI
SETTI

Big urban centres demanded food, fuel and timber. Lots of it. Pressure on the peasants built up and consequently land-care often went out of the window. Nature was abused and misused.

Civilisations collapse

Ecosystems are self-healing – but only up to a point. Failure to take this point felled many a grand culture over the centuries.

The Bible blamed it on God's wrath. Similar floods created similar myths in the Americas, Polynesia and India.

Breaches of the Eco-Code would in time cut spectacular swathes of erosion through China. While in Indochina . . .

Of course these were mere primitive societies compared with the Greeks & Romans.

OK, social justice wasn't their hallmark. But surely they understood that soil is the great bridge between the inanimate and the living?

ARISTOTLE

While the First Democracy's soi! was blowing into the sea the First Republic was building the first city of a million inhabitants – and the world's biggest sewer.

Brilliant Roman engineering carried human waste out of sight . . . but just dumped it in waterways on the outskirts. At the same time burial consisted of tossing bodies into open pits ringing the city.

Rome became a stinking metropolis, ravaged by plague, and in the words of one historian *'recorded a low point in sanitation and hygiene that more primitive communities never descended to'.*

DILUTION IS THE SOLUTION TO POLLUTION

THEN WHY IS EVERYONE DYING ?

The Romans had other ways of killing themselves. To avoid copper poisoning they made their eating and drinking vessels out of lead.

LEAD POISONING TASTES NICER

AND THE SYMPTOMS ARE MORE SUBTLE

YOU GET STUPID

... AND IRRITABLE AND AGGRESSIVE

AND DEAD

The Christians decided that nature had no reason for existence but to serve humanity.

Enough Christians remained uneaten to start the Church – and put a stop to the widespread belief that every tree, stream, hill or any other natural object had its own guardian spirit. The Church destroyed this ungodly 'animism' so thoroughly that by the Middle Ages there weren't many trees left in Europe.

ST FRANCIS

Christian radical. Preached equality of all species in the biosphere. Booted out of the Church. Proclaimed Patron Saint of Ecology by the Pope in 1980.

In the city the air was thickening. Edward I of England decided that execution was the solution to pollution.

SO HOW DO WE KEEP WARM?

WE COULD BURN RATS

On Payne of Death the Burninge of Coal be henceforthe forbydden by law in London Towne that the healthe of the Knights of the Shire may not suffere during theyre resydence herein.

Edward I Rex 1322

Those who didn't freeze to death, choke or hang were treated to the Black Death, an epidemic of bubonic plague that devastated Europe in the 14th century.

AGRICULTURE WAS RUN DOWN EVERYWHERE

IT'S GREAT TO BE A WEED THESE DAYS

WE MAY BE SMALL...

...BUT WE KILLED ONE THIRD OF THE POPULATION

SKRITCH SCRATCH

The Middle Ages brought science & technology to Europe. They also brought feudal ownership of land by a ruling class of kings & queens, nobles and churchmen. Nature was not the only one being exploited.

Peasants, labourers and craftsfolk were doing all the producing. But who was taking home the winnings?

24

These landowners were more concerned with defending their own positions of power than with such trivia as long-term soil fertility. The merchants who succeeded them couldn't tell an earthworm from an eagle. They were just interested in money. But by the end of the Middle Ages they were getting alarmed.

Spilling out over the map, the European states and their enterprising merchants discovered rich new worlds and turned them into poor ones. Spain and Portugal set the tone, conquering, murdering, looting and enslaving across the globe. While England, France and Holland were messing around in the Indies, the Spanish by mistake 'discovered' America . . .

25

America clearly had to be saved from itself. The place was in a chaotic state. Just listen to an eye-witness account:

' *This is a nation in which there is no kind of commerce, no knowledge of letters, no science of numbers, no title of magistrate or of political superior, no habit of service, riches or poverty, no contracts, no inheritance, no divisions of property, only leisurely occupations, no respect for any kinship but the common ties, no clothes, no agriculture, no metals, no use of corn or wine. The very words denoting lying, treason, deceit, greed, envy, slander and forgiveness have never been heard . . . a sick person is a rare sight.* '

Michel de Montaigne

In contrast to these 'savages' the Incas did have agriculture. Describing it, a 20th-century soil expert would say "there has probably never been a system in which agricultural practice and soil organisation were so locked together in a perfect artifact of the mind and spirit".
The Conquistadores swiftly wrecked it.

The Spaniards, having ruined much of their own agricultural land by deforestation and overgrazing, introduced both to the New World.

The plucky European adventurers took home not only gold and silver bullion but also more earthy things . . .

Parmentier introduced the American potato to France by posting an armed guard round the potato patch during the day and withdrawing it at night.

But the new crops that appealed most to the big powers were those that provided staple foods and created the basis for *plantation* economies . . .

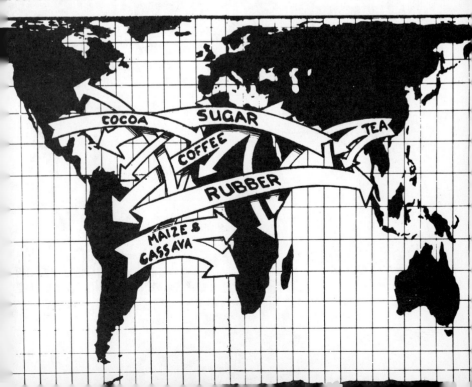

Revolutions

Britain more or less took over in the 18th century. British ships dominated the sea routes and therefore world trade. British capital mounted up. And the British bourgeoisie, the new middle-class rulers, decided the most profitable course would be an Industrial Revolution.

The money was there, thanks to the looting of India, and the machines of mass production were soon invented. But where was the labour?

Scientific, systematic farming with new machines and methods reduced the labour force. Violent and fraudulent land seizures completed the job. Kicked off the land, most peasants had no choice but to become factory fodder – overworked, underpaid and undernourished, reduced from creative workers to 'a mere appendage of flesh on a machine of iron'. They were joined by many craftsfolk whose livelihood had been undercut by machine-made goods.

WANTED DEAD OR ALIVE
for wrecking machinery
and inciting riots

The Art of Housekeeping

Every human society depends for
support on a *resource base*
(energy and materials).

Primitive communities, slave
states and feudal societies relied
mainly on *renewable* energy –
from the sun, wind, and water –
and materials that could be
replenished.

They made little lasting impact on
nature because their production
was restricted by nature's own
productivity.

Until capitalism set about
industrialising society, nearly
80% of all the things used by
humanity came from the animal
kingdom and 20% from the
mineral kingdom.

Industrialism reversed the trend,
concentrating on the Earth's store
of *non-renewable* energy and
materials, such as fossil fuels and
mineral ores.

JUST
1/8 TH OF
SECOND
AGO

THE WORD ECOLOGY COMES FROM THE WORD OIKOS

MEANING HOUSE OR HABITAT

The enclosure of common land in Britain signalled the shift from feudalism to capitalism. It also marked the end of a collective view of the resource base and the beginning of a private, egotistical view.

Industrialism switched the emphasis from *reproductive* use of the resource base, which leaves it intact, to *extractive* use, which reduces the total store. Humanity began draining the Earth's 'capital' instead of living off the 'interest'.

FOOLISH HOUSEKEEPING IN ANY HABITAT

HISTORICAL
CALENDAR

Industrial capitalism launched the Modern Age, and ripped into the available raw materials with no regard for environmental consequences. By the 19th century Britain had become the 'workshop of the world'.

Having exhausted the supply of fresh trees the industrialists solved the fuel shortage with *fossilised* trees – coal.

The steam-engine churned. The cotton-mills hummed. The iron industry boomed. New coal-pits were opened. Towns grew into cities and village workshops into factories.

Canals, roads and railway lines criss-crossed the country. Britain throbbed with industrial activity, to quote the school history books. But what else happened?

Industrialism

Factory fumes and waste poisoned the air, water and soil. The chemicals that spewed into the environment hit hardest at the working classes living nearby, often in overcrowded, diseased slums. The factory owners could afford to live away from ugly sights, sounds and smells.

THIS WAS THE SCENE AROUND THE ALKALI FACTORIES

The foul gases, which belch forth night and day from the many factories, rot the clothes, the teeth and in the end the bodies of the workers, and have killed every tree and every blade of grass for miles around.
— Contemporary account

Marx's partner, Friedrich Engels, crossing a Lancashire river, spoke of *"the most disgusting blackish-green slime pools from the depths of which bubbles of miasmatic gas constantly arise and give forth a stench unendurable even on the bridge 40 or 50 feet above the surface.'*

Marx records that the factory owners resisted all measures for maintaining cleanliness and health if they cut into profits.

SMART BLOKE, OLD MARX

WELL... ER... IN MY DAY PERHAPS

ENGELS

MARX

Marx and Engels, living in the 19th century, saw and wrote about the environmental effects of capitalism. But they weren't too upset about the damage done to nature. In that age of boundless technological optimism *ecological principles were scarcely known — especially the basic idea that the Earth's resources are limited.

Marx seemed to view nature as a constant, static element – the unchanging background against which the class struggle would be fought out.

All human cultures encroach upon nature, altering their environments. But some act as if they owned it instead of being its trustees. Would Marx have revised his view of nature's role had he known that a century later environmental destruction would have reached a point where it threatened the very survival of humanity?

* The word 'œcology' was coined by the German Ernst Haeckel in 1866. It appeared in English as 'ecology' in 1893.

NEXT QUESTION PLEASE...

Some ecologists blame the Baconian creed . . .

WHAT'S THE BACONIAN CREED?

SCIENTIFIC KNOWLEDGE IS TECHNOLOGICAL POWER OVER NATURE

Francis Bacon, a 17th century British Lord Chancellor, provided the industrialists with a scientific alibi for their ruthless behaviour. Jailed for taking bribes, he eventually died of pneumonia while experimenting with frozen chicken.

WELL, AT LEAST THE EXPERIMENT WORKED

The Industrial Revolution stripped the Earth of its stored riches at an alarming rate, broke links with the soil (forcing Britain to import wheat) and pushed the farmer into the background. Populations exploded as technology 'mastered' the environment. Material wealth increased, at least in the West. But at what price?

'IF DIRECTED BY IGNORANCE, WEALTH IS A GREATER EVIL THAN POVERTY, BECAUSE IT CAN PUSH THINGS MORE STRONGLY IN THE WRONG DIRECTION.'

PLATO

Industrialism spread from Britain to Europe and the United States.

Before the 17th century wealth per head in many parts of India, China, Africa and America was *higher* than in Europe.

European capitalism changed the whole picture.

The Trade Triangle

... AND THEN BY ECONOMIC EXPLOITATION

Europe's industrialists turned the rest of the world into suppliers of food, raw materials and labour. Or markets for European manufactured goods.

OR BOTH

In 1660 the French traveller Bernier found Bengal (then part of India) to be richer even than Egypt. He wrote: *'It exports in abundance cottons and silks, rice, sugar and butter. It produces amply for its own consumption of wheat, vegetables, grains, fowls, ducks and geese. It has immense herds of pigs and flocks of sheep and goats. Fish of every kind it has in profusion.'*

In 1757 the British soldier Clive reported Bengal's capital Murshidabad to be *'as extensive, populous and rich as the City of London'*.

Just 30 years of colonial rule later a British MP wrote of India: *'Many parts of these countries have been reduced to the appearance of a desert. The fields are no longer cultivated; extensive tracts are already overgrown with thickets; the husbandman is plundered; the manufacturer oppressed; famine has been repeatedly endured, and depopulation has ensued.'*

Self-sufficiency was destroyed, leaving the colonies at the mercy of the world capitalist market. Mixed farming for subsistence was replaced by plantation farming of single 'cash crops' over vast areas – *monoculture*!

PROVIDENCE SENT THE POTATO BLIGHT ...

To the British ruling classes Ireland was just a source of cheap food — a wheat-growing colony. The Irish peasant's earnings were so small after rents and taxes that he had to grow potatoes to feed his family. Between 1700 and 1845 Ireland went from a grain-based to a potato-based agriculture. In 1845 the potato crop failed

Starvation and hunger typhus killed two million Irish in four years, a quarter of the population. Two million more had to emigrate to the U and Canada. Many them died en route what came to be called the 'coffin ships'.

... BUT ENGLAND MADE THE FAMINE

Colonialist famine

In London the politicians were terribly concerned . . .
. . . about disrupting the grain market . . .

In 1847, when hundreds of
thousands were dying, food worth
17 million pounds was *exported*
from Ireland under the protection
of British troops.

The Irish were killed not just by
monoculture and famine but by
rent, profit and economic theory.

The Irish arrived in the US just in time to witness the annihilation of the greatest animal gathering in history – the buffalo herds of the Western Plains. The American Indian culture almost followed the buffalo into extinction.

THEY PROVIDED US WITH THE NECESSITIES OF LIFE

I KILLED 4,280 BUFFA IN ONE YEAR

Buffalo Bill

Tepees, food, clothing, bedding, fuel, bowstrings, glue, thread, cord, rope, saddle coverings, water vessels, boats and the means of purchasing all they wanted from the traders!

Systematic slaughter between 1850 and 1883 left only a few buffalo placed under protection as an embarrassed afterthought.

WE VIEWED IT AS AN ENEMY TO OVERCO.

RESERVATION

WE TREATED NATURE WITH HUMILITY AND RESPECT

Starting with George Washington in 1779 the US government had long used environmental destruction as a military weapon to subdue the Indians. At exactly the time he was planning to ruin the Iroquois Nation's crops in northern New York honest George declared:

'The more I am acquainted with agricultural affairs the better I am pleased with them . . . I am led to reflect how much more delightful to an undebauched mind is the task of making improvements on the earth than all the vain glory which can be acquired from ravaging it'

The US Army consistently attacked the Indians' natural resources. Between 1860 and 1864 it wiped out the Navahos as a functioning society by destroying all their livestock, orchards and crops.

Eco-warfare was of course nothing new . . .

WE OFTEN BURNED ATHENIAN GRAIN CROPS

WE WRECKED THE TIGRIS IRRIGATION WORKS IN MESOPOTAMIA

WE STOPPED THE FRENCH BY FLOODING OUR OWN LANDS

AND THE RUSSIANS TRIED TO STOP ME BY SCORCHING THEIR OWN EARTH

CAN'T WAIT TO GET AT THOSE OIL WELLS

SPARTAN   MONGOL   DUTCH   BONEY   SADDAM

Building empires

Towards the end of the 19th century the Golden Age of Industrialisation had used up so many raw materials and piled up so many goods that the need for new resources and markets became desperate. So while the US was grabbing colonies around the Americas, the Europeans carved up Africa among themselves.

In India the British had tried to impose their own ideas and practices, building giant canal, road and railway networks, and what one Englishman called 'the whole paraphernalia of a great civilised administration'.

The result? Serious damage to the environment and severe nutritional problems for the rural masses, most of whom grew poorer.

Now it was the turn of Darkest Africa to be enlightened.

World War I was, among other things, a battle for resources between the empire-builders. At stake was the iron-rich Lorraine region (now part of France) as well as colonies and spheres of influence in Africa, eastern Asia and the Pacific. Over 15 million people were killed. In Europe, huge areas of farmland and forest were laid waste, especially in France and Belgium. Nature was also introduced to the delights of chemical warfare.

THE CHEMICAL INDUSTRY AND I JUST DON'T GET ON

Pressing ahead with the modernisation of 'backward' regions British engineers designed the first Aswan Dam on the River Nile in Egypt, promoting an epidemic of the parasitic disease *bilharzia*.

In the three years after the dam's completion in 1919 the disease increased five-fold. Today more than 50% of the population in five African nations have *bilharzia* or similar diseases. In some irrigated areas 100% of the local inhabitants are infected.

LOOK, I SAY, THE DAM THING TURNED DRYLANDS INTO FARMLANDS, DIDN'T IT?

UPSTREAM, YES, BUT DOWNSTREAM IT MESSED UP FARMING AND FISHING

EVER HEARD OF AN ECOSYSTEM?

BRITISH ENGINEER

FRESHWATER SNAIL (BILHARZIA CARRIER)

43

Meanwhile, the industrialists had come up with an impressive new way of poisoning the environment – mass motoring.

The petrol-powered automobile, symbol of the postwar economic boom, crashed into pedestrians, cyclists, horses, dogs, cats, trees, rivers, lamp-posts, shopfronts and, mostly, into other petrol-powered automobiles. The Great Crash, however, was not a traffic accident but a slump which depressed Western economies for years.

Life was particularly depressing for the American farmer. He was losing his soil. Erosion, caused mainly by monoculture (grain), artificial fertiliser, rash tree-cutting and ploughing up too much grassland, turned large regions of Oklahoma, Texas and Kansas into desert during the 1930s. It was called the Dust Bowl disaster. Credit banking played its part by forcing the farmers to over-exploit soil in order to meet interest payments.

44

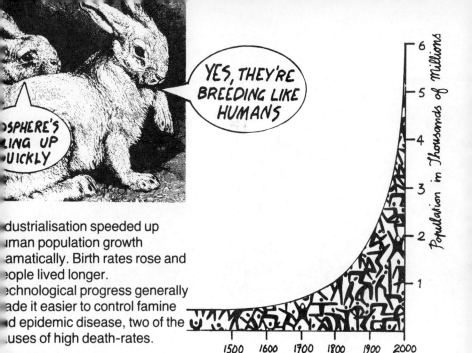

dustrialisation speeded up
uman population growth
amatically. Birth rates rose and
ople lived longer.
chnological progress generally
ade it easier to control famine
d epidemic disease, two of the
uses of high death-rates.

owever, technology did not
me to grips with the *third* cause
uman aggression. Instead it
creased man's destructive
iciency.

orld War II cut the population by
million and wrecked cities,
riculture, forestry and
osystems around the planet.
e of Nazi Germany's
tifications for waging war was
ck of living space' . . .

To cap everything, *Homo sapiens* acquired the power that fires the sun and put it to use in the service of humanity...

Thus human civilisation in the mid-20th century reached a point where single individual or small group could trigger immeasurable catastrophe affecting all life on earth.

Environmental destruction

is any human activity that worsens the prospect for present or future generations to avail themselves of nature and survive within it. From that point of view the difference between agricultural societies and industrial societies is the difference between a light breeze and a storm. And comparing the first half of the 20th century with the last half is like comparing . . .

... A COUGH WITH A DEATH RATTLE !

Post-war development was dedicated to the idea of unbridled technology and unlimited production and consumption. It fuelled the capitalist West and communist Eastern Europe alike. They disagreed on who should run things but shared the same production *goals* and *methods*, viewing nature as infinite and waste as no problem. So they caused the same kind of damage.

THERE WERE DIFFERENCES, THOUGH

IN THE WEST YOU COULDN'T HIDE THE DAMAGE

TOO MANY ENVIRONMENT-ALISTS

WHILE SOVIET COWS...

... HAD NO VOICE

Spreading lead over devastated reactor.

The Chernobyl nuclear power station disaster was a case in point.

A reactor meltdown released up to 200 times the combined radiation of Hiroshima and Nagasaki, and some 400,000 people were forced to leave their homes in the region. Deaths and cancer cases could not be properly determined, as people scattered to all parts of the former Soviet Union.

Until then, millions of East Europeans had no idea nuclear power was even dangerous — like many Americans before the Harrisburg accident of 1979.

Eastern Europe

With the collapse of communism, however, Eastern Europe was revealed as a disaster area — by ecological, economic or any other standards. Seventy years of dictatorship had left whole regions slagged and sludged, swamped and burned, polluted, eroded and salinated.

Marx was confined to the trash heap of history.

Environmental destruction in other parts of the industrialised world seemed almost modest by comparison.

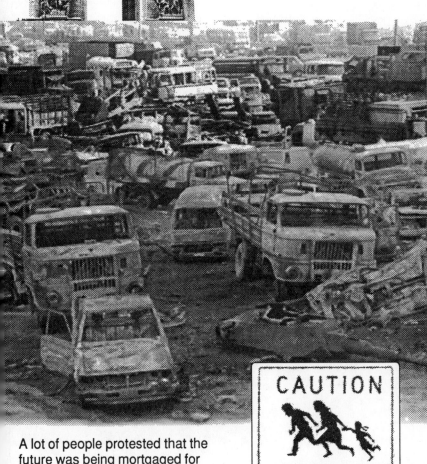

CAUTION

A lot of people protested that the future was being mortgaged for the sake of short-term profit. But they weren't getting much intellectual leadership from those in a position to sway public opinion...

Protects
the car
against the
human rac

Mot
sup

because much of our time is spent working to buy time saving gadgets to do time consum

HIGH TIME FOR A BIT OF BIOLOGY

CARS DESERVE LOVING CARE, SAYS POPE

A car, according to the Pope, deserves the same loving care as the human soul.

During a visit to the Vatican's private garage yesterday, he told the Holy See's 40 drivers: "Your profession as chauffeurs should remind you continually that we are all on the road, heading at high speed towards eternity.

not have time for because much of our time is spent working

Photosynthesis

Homo sapiens means Wise Man. It might have been wiser to have left the atom unsplit and let the sun provide the energy – as it had been doing extremely well for millions of years through *photosynthesis*.

FAR TOO PRIMITIVE

SUNSHINE

CARBONDIOXIDE

WATER

OXYGEN WE BREATHE

FOOD WE EAT

The plants breathe in carbon dioxide, soak up the sun, draw water through their roots, breathe out oxygen and feed us. All life depends on greens.

Some scientists view this dependence as a problem and are trying to free humanity from 'green slavery' by the use of advanced technology (biotechnics).

Cycles

Green plants get their nutrition from the various 'cycles' of the biosphere, circulating energy and matter. Everything goes round and round, nothing is lost. Humans are the only species in natural history to produce things that cannot be reabsorbed. The rest of nature completely converts 'waste' into resources.

THE HYDROLOGICAL CYCLE

Less than 1% of the world's water is fresh water. It is scarce in most poor countries and fast becoming scarce in several industrialised ones, including the US.

Water shortages pit communities against one another in California and countries against one another in the Middle East.

Examples of water use:

1 kilo of dry wheat — 1,500 litres
1 litre of milk — 4,000 litres
1 kilo of meat — 20-60,000 litres
1 car — 400,000 litres

The Carbon Cycle

All living things need carbon. In fact they're made from it. And so are a lot of dead things like coal, oil and diamonds.

Deforestation, pollution of the seas and heavy burning of fossil fuels are disturbing the flow of carbon through the cycle.

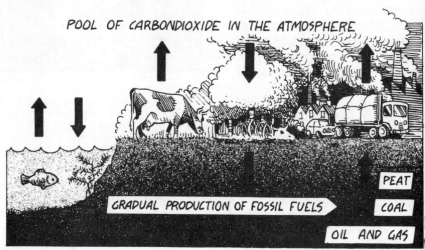

POOL OF CARBONDIOXIDE IN THE ATMOSPHERE

GRADUAL PRODUCTION OF FOSSIL FUELS

PEAT

COAL

OIL AND GAS

Most carbon stays locked in the Earth's rock for millennia. We are releasing it faster by extracting and burning coal, oil and gas. And deforestation is freeing the carbon locked in trees much faster than normal. The result in both cases is a lot of extra carbon dioxide in the atmosphere. The atmosphere acts as a greenhouse for the Earth, letting through incoming light but trapping heat.

An increased amount of carbon dioxide, CFCs, methane and other 'greenhouse gases' causes the atmosphere to block more heat, disrupting the global climate. Food belts could be robbed of rain. If temperatures go up two or three degrees, the polar ice caps may partially melt, raising sea-levels and flooding low-lying areas.

HOLLAND, HERE I COME!

The Greenhouse Effect

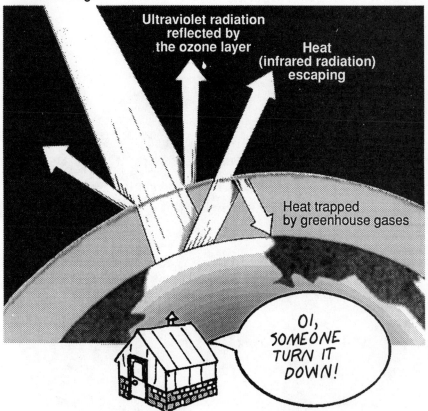

Solar radiation, heat and light

Ultraviolet radiation reflected by the ozone layer

Heat (infrared radiation) escaping

Heat trapped by greenhouse gases

OI, SOMEONE TURN IT DOWN!

Methane is another gas in the global carbon cycle. Although it contributes less to the atmospheric warm-up, emissions are increasing much faster. And each methane molecule has the same greenhouse effect as 30 carbon dioxide molecules. Methane comes from agriculture, especially rice paddies and cattle manure, and also leaks from coalmines and rubbish dumps.

Ozone

A layer of ozone in the upper atmosphere stops us all tanning to a cinder. It protects the surface of the Earth from harmful ultraviolet (UV) rays while letting in enough light to support the growth of plants that form the basis of our food chains. But it is a thin, fragile screen and it is getting thinner at an alarming rate. The first hole appeared over the South Pole in 1985, and kept widening every spring.

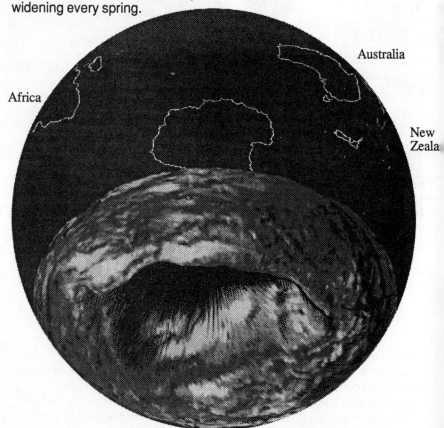

Africa

Australia

New Zeala

A computer-generated simulation of the ozone hole over the Antarctic.

The major causes of ozone depletion are man-made chlorine and bromine compounds, notably the CFCs (chlorofluorocarbons) and halons. When CFC molecules arrive in the upper atmosphere, UV-radiation from the sun splits off their chlorine atoms. Each chlorine atom can then destroy up to 100,000 ozone molecules.

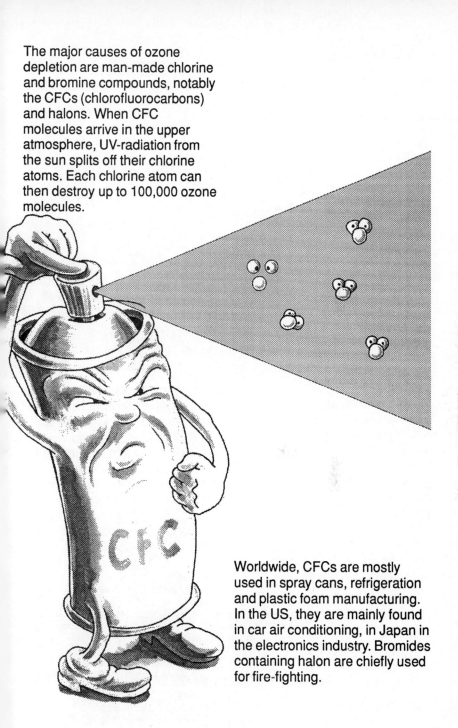

Worldwide, CFCs are mostly used in spray cans, refrigeration and plastic foam manufacturing. In the US, they are mainly found in car air conditioning, in Japan in the electronics industry. Bromides containing halon are chiefly used for fire-fighting.

Acid rain

Descending from the upper atmosphere to the lower, the picture scarcely improves.

Acid deposits degrade biological life in soil and water. As a lake grows more acid, its inhabitants die, starting with the most sensitive. Acidity problems begin at around pH level 6.0. At around pH 5.0, the number of species has shrunk considerably and very few fish are able to survive. Groundwater, too, is threatened. Acid water corrodes drinking-water pipes, releasing such potential health hazards as copper and lead.

At first, soil acidification seems to have a beneficial effect on plant growth by freeing nutrients. But as acidity increases, vital nutrients are leached out of the soil instead and become inaccessible, while aluminium and other toxic metals are released and can be absorbed, threatening tree roots and food chains.

Acidification is primarily the result of sulphur and nitrogen oxide emissions from big heating plants, motor traffic, the pulp and paper industry and smelters. These air contaminants can be dispersed over great distances by winds, coming down either as dry particles or gases or as rain or snow. Locally, forestry and farming practices also play a part.

HEY, LIMING CAN STOP IT!

GOOD, THEN I CAN GO ON POLLUTING

Industry

Areas with plenty of lime in the soil or bedrock are much less vulnerable to acidification. But this leaves many parts of Europe and North America, where the worst damage is to be found. Spreading lime is an expensive solution that has to be constantly repeated. It works better in water than on land. Forest liming takes 30 years to achieve full effect. Farmland recovers much quicker. The best alternative, of course, is to stop the pollution at source...

Eutrophying the waters

Overloading seas, lakes and waterways with nutrients such as nitrogen compounds and phosphates leads to eutrophification. In eutrophied waters, algae and weeds run riot, using up so much oxygen that few species of fish can survive.

Eutrophication is mainly caused by the over-use of chemical fertiliser which leads to fertiliser runoff — the nutrients not absorbed by the plants leak from the soil into adjoining waters. The other big villain is nitrogen dioxide from motor traffic.

The Nitrogen Fix

Chemical fertiliser, mainly composed of nitrogen, is an expensive habit and hard to shake. It can increase yields in the short term but besides disrupting neighbouring ecosystems (and using up valuable non-renewable resources like phosphate) it promotes erosion by allowing the farmer to neglect soil maintenance.

Fertiliser doesn't have to be chemical.
Natural substances you can mix with the soil to stimulate plant growth include manure, rotted vegetable waste, bonemeal, fishmeal, seaweed, powdered lime and sulphur. You can also conjure nitrogen fertiliser out of the air by planting legumes — peas, beans, clover, etc, that draw in or 'fix' atmospheric nitrogen directly in the soil, which is then ploughed.

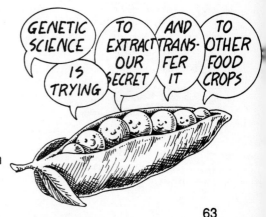

The Healthy Food Chain

What happens to the clean-living, independent plant in a functioning ecosystem? The average blade of grass grows, using nutrients from the soil microbes, dies and rots, and is broken down by the microbes . . .

Cremating human bodies, or sealing them in graveyard coffins, means that the fertility which the blade of grass 'borrowed' from the soil is lost. The chain is broken.

Web, niche & pyramid

SOUNDS LIKE A LAW FIRM

All food chains start with plants. Usually there are only two or three links and almost never more than four, including the *herbivore* (plant-eater) and *carnivore* (flesh-eater).

There are millions of different chains. The sum total of chains in any community is called the *food web*.

Complicated? This is only *part* of food web

Well before the ecologists arrived on the scene Darwin had described nature as 'a web of complex relations'. But he saw life as a competitive struggle for existence won by the strongest of the species. His theory of 'natural selection' delighted the aggressive Victorian capitalists. But a century later it would give way to a view of nature stressing interaction and cooperation, cycles and energy transfers.

MY IDEAS EVOLVED IN INDUSTRI CLIMATE

Darwin showed that every species had its 'place' in nature. This came to be termed the *ecological niche*. It may be filled by different species in different areas. Grass-eaters may be kangaroos in Australia and cattle on the Argentine pampas.

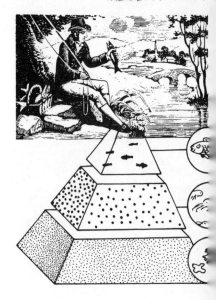

In general the higher up the food chain you go the bigger and fewer the species.

This is the *pyramid* concept. We humans sit at the top of many food pyramids.

66

The Unhealthy Food Chain

WHAT HAPPENS IF YOU SPRAY PLANTS?

Chemical pesticides like DDT do a more thorough job on the predators than on the pests! The pests are more numerous so they adapt. The predators, their *natural* enemies, are vulnerable because the farther up the food chain the DDT travels the more it is concentrated.

THAT'S WHY I'M SO RARE THESE DAYS

Pesticides by removing the biological checks can even create new pests. Once-harmless insects breed out of control when the next ones up in the pecking order are removed.

SPIDER MITE

YIPPEE!! PEST STATUS AT LAST!

Like artificial fertiliser, pesticides lead farmers to greater and greater chemical dependency, upset ecosystems and spread through food chains, joining many other varieties of poisonous waste in human bodies. Their interaction is largely unknown.

DDT is banned in most industrialised countries but its *overall* use is still increasing and the effects are spread globally by wind and water.

AH WELL! EVERY CLOUD HAS A SILVER LINING ... AT LEAST FOR SOME OF US!

67

Stable

A gram of fertile soil contains about 100 million living bacteria. A square metre of farmland can be inhabited by 55 million worms and 50,000 small insects and mites. The total mass of microbe life on the planet has been estimated at 25 times the total mass of all animal life.

The chemical farming industry isn't interested in these little creatures. For quick profit, it drenches the soil in lethal and persistent poisons, ignoring the long-term effects. Fortunately, its activities no longer pass unnoticed...

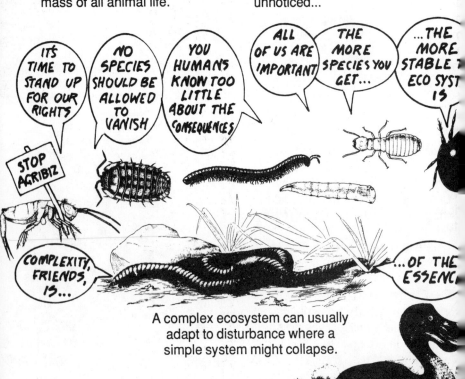

A complex ecosystem can usually adapt to disturbance where a simple system might collapse.

& unstable

Ever since farming began humanity has been an enemy of complexity in nature and therefore a *destabilising* force. Only recently have we begun to realise that losing plant and animal species isn't a trivial matter to be mourned only by nature-lovers but a dangerous and irreversible tinkering with the natural systems on which we depend.

Agribiz is the big culprit. Driven by commercial interest it disregards flavour, food value, resistance to disease, adaptability . . . in short, plant *diversity*. To control the market it's buying up the seed trade and reducing the line to a few patentable, single-season, expensive hybrids dependent on the fertiliser and pesticide which it also sells. To the multinationals biological simplicity makes economic sense. They're filling the world's grainfields with armies of cloned, drugged, sterile zombies requiring constant vigilance and inputs of energy to stave off collapse.

69

Nothing disappears

Freely interpreting the first law of thermodynamics (energy can be neither created nor destroyed), we could say that nothing disappears. The same goes for matter. It just pushes off somewhere else — within our own local ecosystem or into someone else's!

The chemical you pour down your drain, the smoke and fumes from chimneys and cars, the vanishing Aral Sea, the radioactivity from the Nevada tests...it's all kicking around somewhere in the biosphere, in some form (except for some heat that escapes into space).
Not only that, it all spreads — which is the second law of thermodynamics, approximately.

In time, most things turn into something else. They are broken down in nature. But since the dawn of industry we've been producing and spilling things into the environment that can't be broken down. Or that take too long to break down.

Lethal delay

Because many toxic substances move slowly through the environment, or have a cumulative effect, the damage may not appear for a long time. CFCs, for instance, take 15 years to reach the stratosphere.

Acidification and eutrophication sneak up on you. That is why delay in legislating against hazardous materials and practices can be lethal.
All too often, the burden of proof is on the objector rather than the industrialist.

Time for special eco-courts?

Day in the Life of the Lees

DORIS, 47, ELECTRONICS FACTORY HAND

ARTHUR, 45, EX-CAR MECHANIC

GREG, 22, OFFICE WORKER

JULIE, 17, SCHOOL-GIRL

A disturbing tale of a modern nuclear family in contemporary society somewhere in the West.

At home

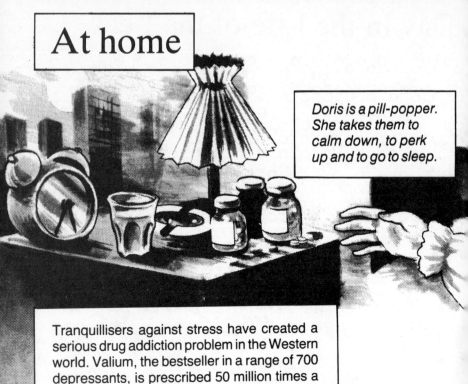

Doris is a pill-popper. She takes them to calm down, to perk up and to go to sleep.

Tranquillisers against stress have created a serious drug addiction problem in the Western world. Valium, the bestseller in a range of 700 depressants, is prescribed 50 million times a year in the US and 20 million times in Europe. Women use twice as much as men. It can be dangerously toxic and produce classic withdrawal symptoms. It earns nearly 1,000 million dollars a year for its makers, Hoffman-LaRoche.

Doris also uses cosmetics although they give her a skin rash. She has seen enough ads to know how the 'ideal woman' is supposed to look. The ad-man 'steals her love of herself as she is and offers it back to her for the price of the product' (John Berger).

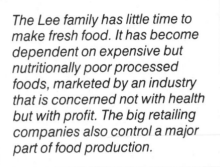

The Lee family has little time to make fresh food. It has become dependent on expensive but nutritionally poor processed foods, marketed by an industry that is concerned not with health but with profit. The big retailing companies also control a major part of food production.

THOSE BATTERY EGGS COST SIX TIMES MORE ENERGY TO PRODUCE THAN THEY GIVE IN FOOD VALUE

Computers have made many work tasks easier, and have greatly improved both communication and access to knowledge via the Internet. But computers can also be addictive and reduce personal interaction. And they help generate massive amounts of random and often misleading information.

BUT WE DON'T EVEN HAVE GAS

GAS BILL
PAY UP OR ELSE
£ 1,000,00

Off to work

The Lees travel a lot. They sleep in one place, work or attend school in another, shop in a third, seek entertainment in a fourth, pursue sports or creative activities in a fifth and 'consume' nature in a sixth.

CAN'T WAIT TO GET A PEDAL CAR

AHH.., FREEDOM

GREG 1

Motor exhaust fumes contain about 1,000 toxic elements and cause 60-90% of all air pollution in industrialised countries. Carbon monoxide is especially dangerous for weak hearts as it reduces oxygen intake. Hydro-carbon causes fatigue and probably cancer. The 8% of humanity who own the world's 400 million cars are also acidifying the rain, warming the Earth and causing 250,000 traffic deaths a year.

OH F....

The automobile is following Greg from the cradle to the grave. He was born in the back seat of a taxi and will die in the front seat of his car with a broken spine and a smashed chassis. Doris goes to work by bus. Julie cycles to school.

HOPE THERE'S ROOM

WISH THERE WERE BICYCLE LANES

In Europe 16 firms are producing seven million cars of 300 different makes every year. Mass motoring drains oil and mineral resources, fouls the air, makes a lot of noise and eats up both land for roads and investment funds badly needed for public transport. The average carrying rate is 1.3 persons per car so it takes 300 cars to transport 400 people – a queue 5 kilometres long.
A train 150 metres long can perform the same task at a fraction of the cost to the economy and the environment.

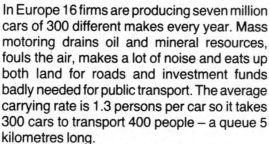

On the job

Doris Lee works on an assembly line. The job is monotonous, requiring little thought or skill, but the fast pace means she can never relax. Doris views the machinery as an enemy. As for the product, she has no idea how it's planned or marketed. She'd like to change her job but doesn't dare because of the high unemployment rate. At the end of the day she rarely has enough energy left to engage in union activity.

Greg Lee works in a factory sales department. His office is clean and quiet. He has never handled a machine and has no idea how the product is put together. Out of touch with the shopfloor, he views the labour force in purely economic terms and agrees with his boss that the unions are troublemakers. Management wants to replace workers with industrial robots and is sending a man to check fully-automated Japanese factories run by a handful of electro-engineers.

Trade unions fight for better pay and working conditions. But what their members gain at the workplace is often lost outside in a deteriorating environment. They're paying more and more to compensate for noise and stress, pollution, inadequate transport, overcrowding, lack of time and lack of access to the countryside. Unions that ignore these 'external' social costs may be ignoring the workers' internal needs and contributing to their alienation.

In hospital

Arthur Lee is in hospital for heart trouble. The place is modern, large, often crowded and short-staffed. Arthur's not sure how ill he is or what treatment he's receiving. No-one has much time for him. He feels small and lonely.

The industrialised countries are getting more and more doctors, more and more hospitals — and more and more sick people. Costs and medicine consumption are going up but there's no corresponding increase in public health and life expectancy.

Aside from AIDS, most deaths are from cancer, heart and respiratory diseases and accidents — usually the result of oppressive environments and lifestyles.

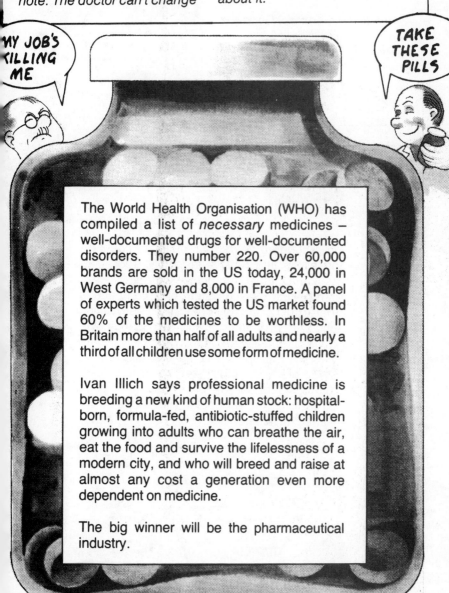

Arthur long ago lost the right to declare himself sick. Only his doctor can do that, with a sick note. The doctor can't change Arthur's social and economic situation but he can give him drugs to stop him worrying about it.

MY JOB'S KILLING ME

TAKE THESE PILLS

The World Health Organisation (WHO) has compiled a list of *necessary* medicines – well-documented drugs for well-documented disorders. They number 220. Over 60,000 brands are sold in the US today, 24,000 in West Germany and 8,000 in France. A panel of experts which tested the US market found 60% of the medicines to be worthless. In Britain more than half of all adults and nearly a third of all children use some form of medicine.

Ivan Illich says professional medicine is breeding a new kind of human stock: hospital-born, formula-fed, antibiotic-stuffed children growing into adults who can breathe the air, eat the food and survive the lifelessness of a modern city, and who will breed and raise at almost any cost a generation even more dependent on medicine.

The big winner will be the pharmaceutical industry.

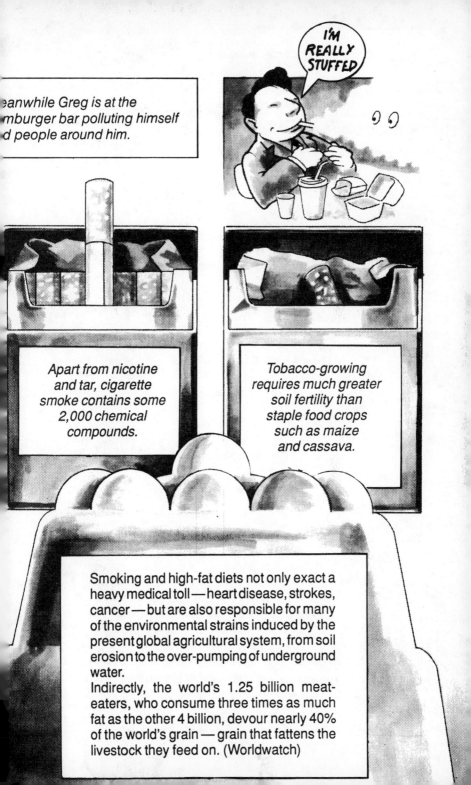

I'M REALLY STUFFED

eanwhile Greg is at the mburger bar polluting himself d people around him.

Apart from nicotine and tar, cigarette smoke contains some 2,000 chemical compounds.

Tobacco-growing requires much greater soil fertility than staple food crops such as maize and cassava.

Smoking and high-fat diets not only exact a heavy medical toll — heart disease, strokes, cancer — but are also responsible for many of the environmental strains induced by the present global agricultural system, from soil erosion to the over-pumping of underground water.

Indirectly, the world's 1.25 billion meat-eaters, who consume three times as much fat as the other 4 billion, devour nearly 40% of the world's grain — grain that fattens the livestock they feed on. (Worldwatch)

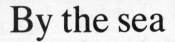

By the sea

Julie Lee is on a school outing.

LOVELY WEATHER!

In places, things are getting a bit *too* hot. The Worldwatch Institute reported in its 'State of the World 2000' survey that global temperatures over the past decade had risen by 0.44 degrees Celsius, and that glaciers were melting from the Peruvian Andes to the Swiss Alps. Global warming is also causing the two ice shelves on either side of the Antarctic peninsula to retreat.

Since 1967, over two million tons of oil have spilled into the seas as a result of tanker accidents, blowouts, wars, etc. Oil often ends up on ocean shores, where most marine life passes a crucial stage of its life cycle.

In the country

Julie Lee's grandparents are buried near the small farm where they worked most of their lives and raised Doris. It has been sold for road and property development.

RIPE
FOR
DEVELOPMENT

Precious agricultural land is being lost at twice the rate that new land is being broken. An area bigger than Britain is disappearing every year. Soil is being exhausted and eroded or it is vanishing beneath motorways, urban spread, airports and industrial development. Or it's becoming wasteland. If this goes on, by the end of the century the world may have to support 1½ times the present population on only ¾ of its present cultivated area.

AND I'M BEING DISPLACED TOO

YESTERDAY'S FARMWORKER IS TODAY'S CANNER, TRACTOR MECHANIC OR FAST FOOD DELIVERY MAN

AGRI BIZ

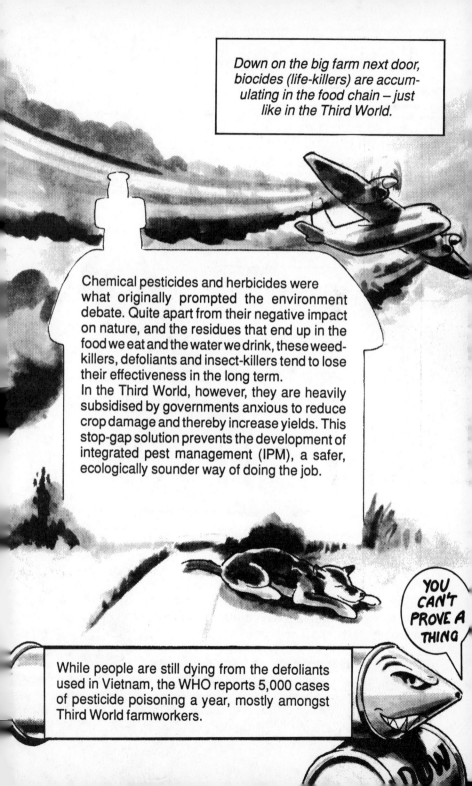

Down on the big farm next door, biocides (life-killers) are accumulating in the food chain – just like in the Third World.

Chemical pesticides and herbicides were what originally prompted the environment debate. Quite apart from their negative impact on nature, and the residues that end up in the food we eat and the water we drink, these weed-killers, defoliants and insect-killers tend to lose their effectiveness in the long term.

In the Third World, however, they are heavily subsidised by governments anxious to reduce crop damage and thereby increase yields. This stop-gap solution prevents the development of integrated pest management (IPM), a safer, ecologically sounder way of doing the job.

YOU CAN'T PROVE A THING

While people are still dying from the defoliants used in Vietnam, the WHO reports 5,000 cases of pesticide poisoning a year, mostly amongst Third World farmworkers.

At the shops

Doris Lee doesn't know the staff at the giant supermarket which has replaced her local shops. To them she is just another face in the crowd. She can afford more in her trolley nowadays but she has a feeling she is not getting more for her money.

Shoddy goods squander energy and raw materials while keeping profits up. Unnecessary packaging, especially throwaway containers, add to the problem – and the profits. The average American throws away over two kilos of household waste a day, the average Briton almost a kilo. Most of it could be recycled.

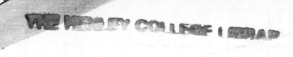

Doris, like most Westerners, spends a third of her income on food. In other words a third of her working life is concerned with getting food – exactly like the hunters and gatherers of early times!

The broiler chicken is the most industrialised of all animal foods. It's a little meat machine, bred to live just six months in a giant factory, sitting in a cage, scarcely moving, almost always eating, under constant medication, pumped with growth hormones and, if it survives the stress, finally slaughtered by an assembly line.

Time for tea

Back home Greg is waiting to be fed.

There is no evidence of a single human behavioural characteristic caused by genetic differences. Ideas about 'social Darwinism' and 'racial hygiene' arose in Germany at the turn of the century and later underpinned the Nazi policy of exterminating Jews and Gypsies. Today they are being used by US and European right-wing groups to justify racism and sexism. Blacks are 'naturally' under-privileged and women are 'naturally' housewives. The poor and starving are being 'weeded out by natural selection', etc, etc. The argument that social differences have a biological basis is convenient for those practising the oppression.

The TV commercials are also teaching comparison. Julie is being urged to keep up, to strive for the better life of glamour and luxury.

BATHE IN THE CHAMPAGNE SURF

The consumer society is always raising the stakes. It deals in frustration by constantly heightening expectations. As soon as something is available to all, it is 'no longer attractive'.
If you had a digital video camera yesterday, you must have home cinema equipment today and a virtual reality room tomorrow. New 'demands' are created by producing and promoting new articles and activities that in fact are accessible only to a small minority of consumers. It's called 'modernising poverty'.

AND A GLOBAL CONSUMER SOCIETY WOULD LAY WASTE THE PLANET

The same evening

Doris is daydreaming. She remembers when she and Arthur had the time and energy to cycle into the country, pick berries and lie among the wildflowers. In their old home he enjoyed doing carpentry in the back yard. Since moving to the housing estate the only things he has made with his hands have been matchstick cathedrals.

The world 'is now lurching from one energy crisis to another, threatening at every turn to derail the global economy or disrupt its environmental support systems'. The world is 'likely to be plagued by more frequent and more severe energy crises than ever before'. (Worldwatch)

The World Energy Conference predicts that by 2020 we will be using 75% more energy and that most of it will be supplied by coal, oil and nuclear power. As Chernobyl constantly reminds us, nuclear power is too risky in both human, ecological and economic terms. Oil and coal are chiefly responsible for the 6 billion tons of carbon being added to the atmosphere every year.

The Worldwatch Institute says that over the next 30 years, industrial countries can halve their energy use without upsetting their economies — if they begin moving in the right direction now.

That night

Greg

dreams of a white super-race produced by genetic engineering. So as not to pollute the gene pool non-whites have been exterminated along with the mentally ill, the physically handicapped and the homosexuals. Women are either in bed or in the kitchen.

Arthur

dreams he has stopped drinking and shouting at his family. Since going on to a short working week he has taken up carpentry again at the neighbourhood workshop. He earns quite a lot doing odd jobs and feels useful and creative.

Julie

dreams she's learning how to work with leather and stone at a school where no one's forced to attend lessons, academic certificates don't exist and afternoons are spent outdoors. The Labour Exchange has been turned into a Free Activity Centre.

Doris

dreams she's a vegetable gardener far from noise, stress and pollution. Her work is varied and meaningful. She feels healthy, playing an active role in a community where everyone knows everyone else. She has stopped taking pills.

The Lees live in a consumerist, technocratic society that makes little sense in human or ecological terms. It destroys their physical environment and their mental well-being.

We can now reach all parts of the biosphere, from the deepest oceans to the farthest reaches of the sky, and communicate almost instantly. But our failure to control technology and treat the land sensibly means that the body fat of every person on the planet now contains DDT and that agriculture, the 'basic interface between human societies and their environment', is rapidly *destroying* soil while millions starve to death.

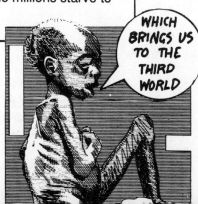

WHICH BRINGS US TO THE THIRD WORLD

Water & Fuel

Life in the Third World is mainly a struggle for survival.

In parts of Africa women today have to walk 15 or 20 kilometres to fetch water, leaving at dusk and returning at dawn. In Africa and South and East Asia only 1/5 of the rural population has reasonable access to a safe water supply.

AH! THIS CALLS FOR FOREIGN AID

AID OFFICIAL

Almost every tubewell wound up as the property of one man – the rich landowner who could bribe the local authorities. He can sell 'his' water at his own price, which is often too high for the other village farmers.

The World Bank lent Bangladesh funds for 3,000 tubewells. Each was supposed to provide irrigation for 25-50 farmers.

So the tubewell is being under-used at the same time as the large landowner is better placed to buy out his poor neighbours when hard times fall.

About half of the wood cut in the Third World is used for fuel.

THIS USED TO TAKE AN HOUR OR TWO, NOW IT TAKES ALL DAY

IN NIGER LABOURERS SPEND A THIRD OF THEIR WAGES ON WOOD

Fuel is so scarce in some African regions that crop residues and stubble are now being burned although they are desperately needed as humus (plant nutrients) to bind the soil.

Indian farmland is likewise losing its fertility because the firewood shortage is forcing peasants to burn cow dung, which now accounts for most domestic fuel consumption!

On Java, forested slopes have been reduced to crumbling cliffs as land hunger pushes peasants higher and higher up mountainsides.

THE POOR MAN'S ENERGY CRISIS HAS COLONIAL ROOTS

Erosion

Desert

Large areas of desert around the world are being degraded by overgrazing, bad farming and deforestation.

While in some countries like Algeria, Israel and China, desert is being restored to farmland and orchards, in general the deserts are advancing, particularly in India and the Sahel, the vast dryland region of Africa along the southern Sahara. Poverty is almost always the cause.

Vulnerable soil is being over-cultivated and erosion is rife. At present, the desert is spreading at the rate of 15-16 million acres a year and the UN Environment Programme estimates that a further 50 million acres a year is being rendered unusable. A fifth of the world population stands to lose its means of livelihood.

Until colonialism came, pastoral nomads herding animals around the Sahel fully utilised the semi-arid desert's few resources.

They moved according to the cycles of nature, kept mixed herds to exploit the various ecological niches, swapped meat and dairy products for the farmers' grain and manured their lands.

The French changed all that. Taxes forced the nomads to deal in money and rely on the 'market'. Their pastures were turned over to peanut and cotton production for export, their movements were restricted and the demand for beef made them give up mixed herds for cattle.

To compensate they have built up herds, with the aid of modern medicine, beyond the carrying capacity of the land left to them.

MODERN RANCHING IS THE SOLUTION

AGRI BIZ

No! REUNITE THE FARMER AND THE HERDER

Tropical rain forest is one of the planet's great biological resources. It has developed an enormously efficient ecosystem over 50 million years. But the system is fragile because the topsoil is shallow and the lush growth is largely water, disguising a low level of fertility. It's the dense tree cover that keeps the system going. Without it the soil is easily washed away.

Plant and animal life in tropical rain forests is simply not equipped to deal with sudden and drastic large-scale change. It's no match for legions of giant bulldozers, chainsaws, fire-teams and biocides — the tools being used to destroy the rain forests at a rate of some 100,000 acres a day.

To strip tropical rain forest is to wipe out for ever the end-product of millennia of evolution (the surviving ancestral species from which most of today's 10 million species have descended), to prevent the study of thousands of undescribed plants and animals, to ruin beauty and to endanger the survival of the remaining 'primitive' cultural groups trying to live there.

Jungle

While the primary forest in the industrialised North is still disappearing at an alarming rate, particularly in Canada (British Columbia) and the US, half of the Earth's rain forest has been cleared in just 20 years! Many countries in the South are exhausting their stands through overcutting and are becoming net importers of wood. The Philippines did it in the 'sixties, Indonesia in the 'seventies, Malaysia looks like doing it soon and next it will be the turn of Mexico, some West African states and, unless replanting is stepped up, Brazil.

BRAZIL'S RICH, BUT WE'RE POOR

The Brazilian jungle is one of the world's biggest 'carbon dioxide eaters', offsetting 20-30% of global greenhouse gases. Brazil is making a lot of money out of forest-clearing (and tends to view ecological concern about the jungle as yet another attempt by foreigners to control its assets). But by nutritional standards, the people of 'rich' Brazil are worse off than the people of 'stagnating' Sri Lanka.

The Green Revolution

In many Third World areas the Green Revolution has replaced ecologically-sound mixed crop strains with uniform, short-stemmed hybrids of rice, wheat and maize specially bred to respond to chemical fertiliser, biocides and about three times as much water as their predecessors. They are hypersensitive, totally reliant on their magic ingredients.

Food production is increasing, but so is deprivation and the risk of ecological collapse. The Green Revolution with its greater emphasis on technology has enriched the already well-off while creating unemployment and pushing the peasants and the landless below the poverty line.

Their ability to buy food is reduced. Capacity grows but 'demand' does not. Mountains of grain rot in countries where people go hungry.

104

In the Punjab, India's 'granary', the Green Revolution 'has led to reduced genetic diversity, increased vulnerability to pests, soil erosion, water shortages, reduced soil fertility, micronutrient deficiencies, soil contamination, reduced availability of nutritious food crops for the local population, the displacement of vast numbers of small farmers from their land, rural impoverishment and increased tensions and conflicts. The beneficiaries have been the agrochemical industry, large petrochemical companies, manufacturers of agricultural machinery, dam builders and large landowners.' (Vandana Shiva)

Before the World Bank, Ford Foundation and other philanthropists decided that India needed 'high-yield' hybrids (thus assuring Agribiz of higher fertiliser consumption overseas), India had successfully pursued a policy that involved strengthening the ecological base of agriculture and the self-reliance of peasants.

Cash Crops

Impoverishment of the soil is no natural disaster. It is mainly a result of Western demand for meat and luxury 'colonial goods'.

Cash-cropping monoculture is steadily replacing traditional techniques that protected the soil. Worldwide, almost 640,000 sq. km. are given over to cash crops of little or no nutritional value. Providing what?

Sugar for the sweet Western tooth
Tea & coffee for Western drinkers
Tobacco for Western smokers
Cotton for Western jeans
Cut flowers for Western vases
Peanuts for Western parties
Feed for Western livestock

Ecologically damaging Western practices are not only to be found in agriculture and forestry. Many Third World countries seem intent on making the same industrial mistakes as the West and are adopting big-scale technologies where small or medium-sized ones would be more appropriate and ecologically more manageable.

National elites backed by the multinational corporations seek to introduce standardised mass culture, middle-class lifestyles and values into economies which can afford none of them and traditional societies which cannot survive alongside them. The 'Western way' tends to reduce humanity's cultural diversity.

Malnutrition

The Green Revolution hybrids displace pulses and other protein-rich crops, thereby adding to the malnutrition that's undermining human capacity in the Third World. Protein deficiency impairs brain development in infants and the damage cannot be undone later.

The WHO says that 100 million of the world's 300 million children under five are suffering from protein energy malnutrition. Under-nourished children have smaller heads than normal. Their brains don't fill their skulls properly. Those who survive infancy grow up chronically malnourished and slip easily into a vicious circle of deprivation . . .

The multinationals are flooding the Third World with products that are useless or directly harmful or both. With one hand they steal the protein from the poor and serve it up as steak, bacon, eggs and milk for Western tables. With the other they pump into the Third World things like Ritz crackers, chewing gum, ITT's Hostess Twinkies cupcakes, white bread, Coke, Fanta, Pepsi, Frosted Flakes, as well as high-tar cigarettes and drugs banned in the West.

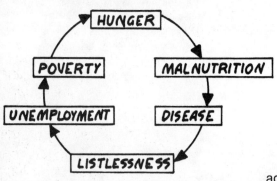

In the Third World 75% of premature deaths are caused by parasitic or infectious disease. In most cases malnutrition has contributed.

Breastmilk is an invaluable source of nutrition. Through aggressive advertising, Nestlé and others have helped to persuade many Third World mothers to use infant formula instead. A major problem is the scarcity of clean water to make it with.

Population

A couple of decades ago, the human population was four billion. It took 100,000 years to reach that figure. Today, there are six billion of us.

I'M THE ONLY ANIMAL WHO CAN KEEP UP

HOW MUCH CAN I TAKE?

No-one is sure of the Earth's maximum 'carrying capacity'. One UN report suggests 36 billion. The US Academy of Sciences has estimated 30 billion. Others put the figure considerably lower.

For the starving villagers of the Third World, where population growth is fastest, the question holds little interest. Children are usually the only resource they can rely on. The rest is owned by the rich.

Birth rates generally do not fall until living standards improve — or until poverty is so great that new children are more of a burden than an asset.

Rising affluence has stopped population growth in many parts of the North. In Europe it is approaching zero. In parts of the South as well — China, Japan, Indonesia — there are signs that growth is slowing.

In much of the Third World, though, especially the Indian sub-continent, rapid population growth is severely overtaxing ecosystems. The natural resource base on which economies rest is shrinking — forest, soil and grasslands can no longer carry the same amount of people, let alone a growing number. The longer it takes to introduce ecologically sound farming and forestry practices along with fair development policies based on local communities, the more terrifying is the prospect for the world's poor.

Family planning has been badly neglected in places, not least because of the withdrawal of US funds from international programmes, exposing more and more women to unsafe abortions as the only alternative to a life of endless pregnancies.

Resources

LET'S GET ONE THING STRAIGHT

There is no global shortage of food.

BUT I THOUGHT...

There is enough food in the world today to feed everyone. The problem is that it's not being shared out fairly.

DO YOU MEAN...?

And that applies to most resources. It's not population pressure that's gobbling them up so much as wasteful production and consumption in the industrialised world.

I REALLY MUST PROTEST!

The North takes 2/3 of world fossil fuel imports, 3/4 of metal ores and 4/5 of non-ferrous metals. North Americans make up 5% of the world population yet consume one third of the planet's resources. Europe accounts for 70% of primary energy and fossil fuel use.

BLOODY SCANDAL!

THERE'S NO GLOBAL LACK...

...OF EARTHLY RESOURCES

BUT THE POOR ARE DENIED ACCESS TO THEM..

IT ALL COMES DOWN TO WHO OWNS WHAT

NORTH Inc.
Debt Collector

Governments and corporations in the rich countries have been so skilful at helping poor countries 'develop' that today more money is flowing from South to North than vice versa. Most of it in the form of debt repayments...

Swapping debts

As long as they are burdened with huge foreign debts, poor countries in the Third World or Eastern or Central Europe cannot afford clean-up or reforestation programmes.

Debt-for-nature swaps are one way out. Governments or institutions in the industrialised world 'buy' poor countries' debts in exchange for the establishment of nature reserves, conservation programmes, etc.

What about the rich countries' ecological debt to the poor? Industrial projects in the North that reinforce the greenhouse effect through carbon dioxide emissions can restore the balance by planting carbon-absorbing forest in a poor country. Thus the carbon score is even.

Best of all is to avoid such industrial projects in the first place.
And to help poor countries with sustainable development programmes of their own — ones that improve the lot of the people as a whole instead of chiefly benefiting consumerist elites anxious to emulate the North.

Scarcity

'If we seek to devise a way of life that can continue, in safety, into the remote future, then we would do well to base that way of life on the use of resources that will remain easily obtainable and plentiful for as long as the Earth continues to exist. Where we are dependent on resources that can be exhausted it would be wise to free ourselves from such dependence.'

— from The Little Green Book

WHAT'S THE DIFFERENCE BETWEEN RESOURCES AND RESERVES?

THE **RESOURCE** IS THE TOTAL GLOBAL STOCK OF SOMETHING. THE **RESERVE** IS THE PART THAT CAN BE EXTRACTED ECONOMICALLY

Resource estimates are notoriously unreliable. They keep changing as new deposits are located. But the general trend is that resources are becoming harder and harder to get at. The oilmen and miners are having to dig deeper or develop new technology, using ever more energy, all of which are expensive.

So even if a resource doesn't vanish completely it's increasingly concentrated in the hands of the rich and powerful, until it's no longer profitable to recover.

In such a 'scarcity economy' life and death can be determined by which class you belong to.

SCARCITY JUST MEANS A RESOURCE CAN'T BE RECOVERED SO PROFITABLY

The flow of the world's resources is dictated by the rich, often through the multinationals. A law unto themselves in the absence of government restraints, the big corporations have built their position — and our dependence on them — on cheap energy and raw materials.

As these become ever scarcer, and the Third World seeks greater control over its fast-disappearing assets, the industrial North may have to choose between loosening its grip on the South or being forced to embark on more and more 'crusades' to protect its interests.

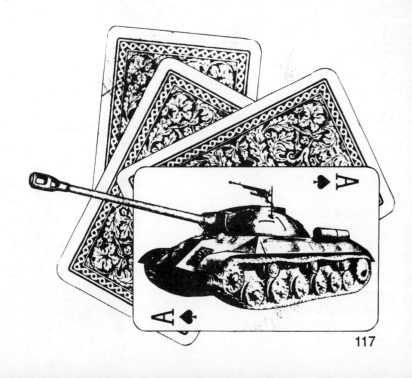

Whose needs?

A lot of people around the planet aren't feeling too well. Their needs aren't being met. What *do* they need?

THINGS ARE LOOKING A BIT GRIM

Protected Species

TO BE LOVED AND TOUCHED

FED CLOTHED AND HOUSED

HEALTH AND EDUCATION

COMPANIONSHIP

ENVIRONMENT

RIGHT AND FREEDO

SURVIVAL JUSTICE AND SELF-DEFINED WORK

What does Illich mean by self-defined work?

'People need not only to obtain things, they need above all the freedom to make things among which they can live, or give shape to them according to their own tastes, and to put them to use in caring about others'
— from Tools for Conviviality

Hi, I'm Ivan Illich

Economists don't concern themselves with our inner feelings, hopes and dreams.

WHAT'S RIGHT FOR THE ECONOMY IS RIGHT FOR HUMANITY

Forced to adapt to a society that measures life in economic and consumerist terms, we easily lose touch with what is good for us. Most psychological data show that above the poverty level the main determinants of happiness are not related to consumption at all.

THE "BRAINS"

'In the final analysis, accepting and living by sufficiency rather than excess offers a return to what is, culturally speaking, the human home: to the ancient order of family, community, good work and good life; to a reverence for the excellence of skilled handiwork; to a true materialism that does not just care **about** things but cares **for** them; to communities worth spending a lifetime in.'

— Alan Durning

DON'T SEE MANY ADVERTS FOR THAT!

119

Which economy?

The 'command economies' of socialist Eastern Europe failed to deliver the goods. The Soviet workers' state could not even keep its miners in soap. Yet in its development view, socialism was no different from Western capitalism. It believed that nature was simply an object to be exploited and that technology could solve all the problems.

In a planned economy like the Soviet one, environment protection could be and was ignored if it got in the way of industrial growth. There being no democracy, there was no grass-roots environment movement to cry wolf. (Eventually, environmentalists were at the forefront of the revolution that ousted communism.)

Is the market economy, then, ecologically-minded? Hardly. Industry does tend to move on when it can no longer dig deeper, build higher or pay more for land and energy without upsetting profits. But by then it may be too late for the environment. So market excesses must be checked by firm government — preferably before they occur!

In a sustainable economy, the environmental costs of any given activity must be fully reflected in the price to the consumer. This applies to everything from the dismantling of clapped-out nuclear reactors to roadbuilding or flying fresh New Zealand fruit to Europe.

In a sustainable economy, 'eco-taxes' would make it more profitable for people to eat food produced closer to home, to commute by bike or bus...

...to buy long-life products...

...to get their energy from the wind and sun...

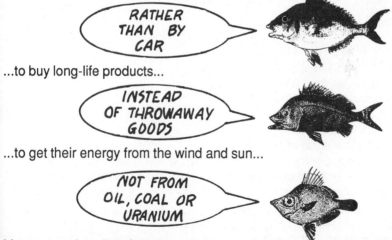

Many countries already have such taxes or levies, e.g. on air and water pollution, fertilisers and batteries. But they are usually set too low.

Growth & misgrowth

Today, economic growth in the North generally means more commercialism and more consumerism, more pollution, more centralisation, more alienation — and less resources for the South, where they are needed most.

SO WHAT'S THE ALTERNATIVE?

If we have reached a point where 'more' often means 'worse' it might be a good idea to re-define growth. It is usually measured in GNP (gross national product). But this is a misleading guide to the ecological state of a nation or the well-being of its people. For instance, GNP goes up when you get more road accidents (hospitals, car-makers, etc, are kept busy) or when goods are so shoddy or hard to mend that people have to keep buying new ones. The economy may hum, but at what cost to our environment and to us?

PERHAPS A RICHER LIFE DEMANDS FEWER PRODUCTS?

AND SELECTIVE GROWTH OF THINGS THAT WE DECIDE!

The other question is, how much can we fairly expect? How much is enough, not just from a personal viewpoint but from a global one? A newly-liberated, democratic Eastern Europe bent on satisfying consumer needs that have never been met represents a colossal market. China and India are moving in the same direction. Why should the citizens of these countries expect less than the high-consumption societies of North America, Japan and Western Europe?

An Indian Speaks

Back in 1980, American Indian Russell Means foresaw the ecological collapse of Eastern Europe — and indeed of Western society.
Excerpts from an address at the Black Hills Alliance Survival Gathering:

*Being is a spiritual proposition. Gaining is a material act. Traditionally, American Indians have always attempted to be the best people they could. Part of that spiritual process was and is to give away wealth, to discard wealth in order **not to gain**. Material gain is an indicator of false status among traditional people while it is 'proof that the system works' to Europeans . . .*

*Most important here is the fact that Europeans feel no sense of loss in all this. After all, their philosophers have despiritualised reality so there is no satisfaction (fcr them) to be gained in simply observing the wonder of a mountain or a lake or a people **in being**. Satisfaction is measured in terms of gaining material – so the mountain becomes gravel and the lake becomes coolant for a factory . . .*

I do not believe that capitalism itself is really responsible for the situation in which (American Indians) have been declared a national sacrifice. No, it is the European tradition. European culture itself is responsible. Marxism is just the latest continuation of this tradition, not a solution to it. There is another way. There is the traditional Lakota way and the ways of the other American Indian peoples. It is the way that knows that humans do not have the right to degrade Mother Earth, that there are forces beyond anything the European mind has conceived, that humans must be in harmony with all relations or the relations will eventually eliminate the disharmony.

All European tradition, Marxism included, has conspired to defy the natural order of all things. Mother Earth has been abused, the powers have been abused, and this cannot go on forever. No theory can alter that simple fact. Mother Earth will retaliate, the whole environment will retaliate, and the abusers will be eliminated. Things come full circle. Back to where it started. That's revolution.

LET'S TRY TO PIN DOWN ECOLOGY

AND SEE WHERE IT COMES INTO THE PICTURE

Ecology is more than just a branch of biology. It brings together a string of natural and social sciences, as well as philosophy, and studies nature *as a whole*. This 'holistic' approach is what makes it such a broad subject. Its central theme is the interdependence of all living things.

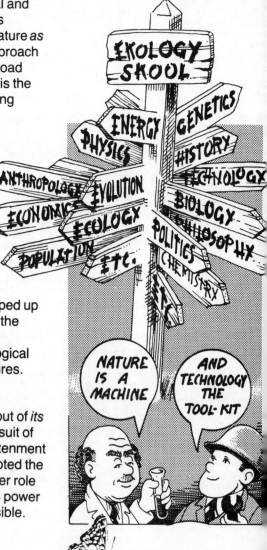

EKOLOGY SKOOL

ENERGY
GENETICS
PHYSICS
HISTORY
ANTHROPOLOGY
TECHNOLOGY
EVOLUTION
BIOLOGY
ECONOMICS
PHILOSOPHY
ECOLOGY
POLITICS
POPULATION
CHEMISTRY
ETC.
ETC.

NATURE IS A MACHINE

AND TECHNOLOGY THE TOOL-KIT

Ideas about ecology cropped up in the 18th century, when the industrial capitalists were breaking out of their ecological niches and founding empires.

Science had just broken out of *its* traditional niche – the pursuit of wisdom for human enlightenment and perfection – and adopted the idea that humanity's proper role on Earth was to extend its power over nature as far as possible.

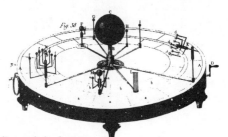

This 'mechanical' model of nature has dominated Western thinking ever since, and served the industrialists well. It has also shaped the development of ecology. Some modern ecologists call their field 'bio-economics' — nature is transformed into a business division.

Right from the outset, however, there were ecologists who insisted that humanity could never be an island unto itself.

I'M FOR PEACEFUL CO-EXISTENCE WITH NATURE

Gilbert White, English Curate, 1720-1793

ALL IS ONE AND INTER-RELATED

Henry David Thoreau, American Naturalist 1817-1862

The 20th century brought a growing number of them. At first such views made little impact on the owners and planners of industrial society or on the public consciousness. But then...

127

The Age of Ecology

Smack in the middle of the West's Second Industrial Revolution, a 1962 book by an American nature writer, Rachel Carson's *Silent Spring*, created a sensation. It showed how the new biocides flooding the environment were a threat to humanity on a par with nuclear war. Carson shared White's and Thoreau's vision of the unity of all life and urged scientists to take a humbler approach to nature.

> WE MUST REALISE WE ARE ONLY A TINY PART OF A VAST AND INCREDIBLE UNIVERSE

The book launched the modern ecology movement and the fires were fuelled by another American biologist, Barry Commoner, who analysed the explosive increase of pollution and found the ecological chains were being broken by industry's new production techniques and its replacement of natural products with synthetics.

> PRODUCTION FOR PROFIT TENDS TO BE HIGHLY DESTRUCTIVE

By the 1970s, the message was loud and unmistakable — there were indeed limits to growth. Graphs showed population and pollution soaring and resources dwindling, and somewhere in the mid-21st century the collapse of the industrial base, taking with it the agricultural and service sectors. Suddenly, the ecologists had everyone's ear...well, almost everyone's...

In the 1980s, the environment lobby swelled from a small community of experts into a grass-roots movement, well-informed and determined. Across the globe, people engaged in a wide variety of local and national actions that pointed towards a safer, more human environment, a fairer world order and a more satisfying lifestyle.

By the 1990s, with bottled fresh water selling well and holes in the sky forever widening, there were green parties and groups in most corners of the world and the science of ecology was firmly rooted at the centre of any serious discussion about desirable futures.

Ecology alone

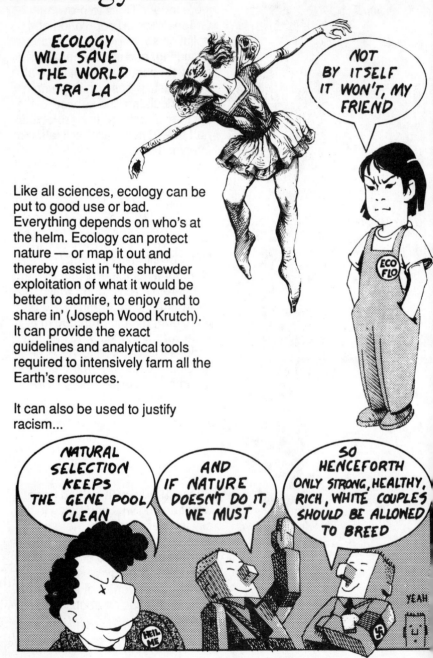

ECOLOGY WILL SAVE THE WORLD TRA-LA

NOT BY ITSELF IT WON'T, MY FRIEND

Like all sciences, ecology can be put to good use or bad. Everything depends on who's at the helm. Ecology can protect nature — or map it out and thereby assist in 'the shrewder exploitation of what it would be better to admire, to enjoy and to share in' (Joseph Wood Krutch). It can provide the exact guidelines and analytical tools required to intensively farm all the Earth's resources.

It can also be used to justify racism...

NATURAL SELECTION KEEPS THE GENE POOL CLEAN

AND IF NATURE DOESN'T DO IT, WE MUST

SO HENCEFORTH ONLY STRONG, HEALTHY, RICH, WHITE COUPLES SHOULD BE ALLOWED TO BREED

YEAH

. . . or to confuse the issue and cement inequality.

WE'RE ALL IN THE SAME BOAT

Understanding the way nature works is vitally important. Ecology can establish what we can and cannot do if the 'web of life' is to be kept intact, and it can be used to radically criticise society. But in itself it is only a tool.

To decide how it should be used means choosing between different lifestyles, systems and types of civilisation.

AND THAT FOLKS IS A POLITICAL CHOICE

Choosing

THESE ARE THE ALTERNATIVES AS ILLICH SEES THEM

Either we agree to impose limits on technology and industrial production so as to conserve natural resources, preserve the ecological balances necessary to life, and favour the development and autonomy of communities and individuals . . .

HE CALLS THIS **CONVIVIAL** SOCIETY

. . . or else the limits necessary to the preservation of life will be centrally determined and planned by ecological engineers, and the programmed production of an 'optimal environment' will be entrusted to centralised institutions and hard technologies.

HE CALLS THAT THE TECHNO-FASCIST PATH

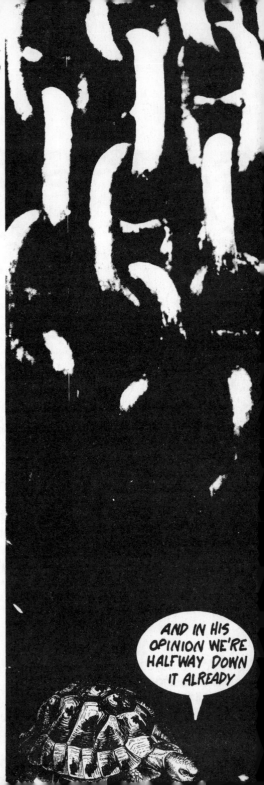

AND IN HIS OPINION WE'RE HALFWAY DOWN IT ALREADY

futures

Technology is supposed to provide for human needs. But its main function has been to substitute non-human energy and machines for people.

Modern technocratic society has developed the kind of technology that serves its interests, and either ignored or obstructed techniques that do not — organic farming, renewable energy systems, etc. This policy has usually been destructive of nature. So scientists are now having to spend whole careers trying to 'copy' nature and create artificial life in laboratories.

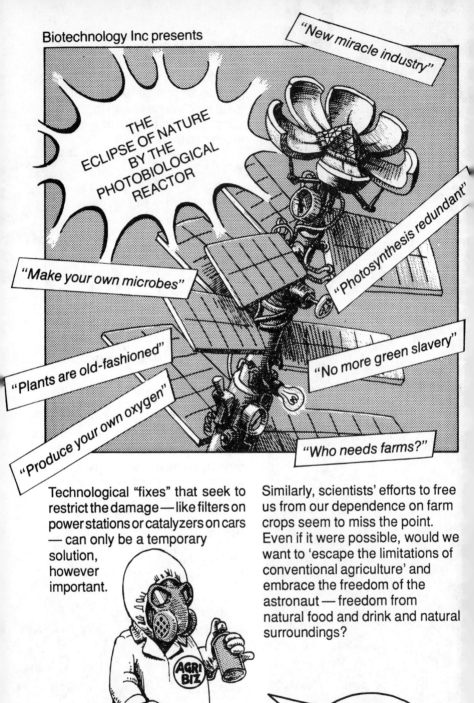

Biotechnology Inc presents

THE ECLIPSE OF NATURE BY THE PHOTOBIOLOGICAL REACTOR

"New miracle industry"

"Make your own microbes"

"Photosynthesis redundant"

"Plants are old-fashioned"

"No more green slavery"

"Produce your own oxygen"

"Who needs farms?"

Technological "fixes" that seek to restrict the damage — like filters on power stations or catalyzers on cars — can only be a temporary solution, however important.

Similarly, scientists' efforts to free us from our dependence on farm crops seem to miss the point. Even if it were possible, would we want to 'escape the limitations of conventional agriculture' and embrace the freedom of the astronaut — freedom from natural food and drink and natural surroundings?

AGRI BIZ

CAN'T IMAGINE WHY NOT

只今の公害状況
オキシダント濃度
0.07 ppm
ここの騒音 72 ホン
東京都

Wouldn't it be wiser to attack pollution at source — to avoid producing it in the first place? To shift from road to rail traffic as far as possible? To shift from fossil fuels to renewable energy, from chemical to organic farming? From wasteful to low-waste or non-waste technologies?

Fortunately, such moves are afoot, here and there. Technology is at last being harnessed to provide a sustainable future. People and governments are aware of the dangers and are doing something about the problems. We're getting some eco-solutions...

135

Goods must be good

OK, LET'S LOOK ON THE BRIGHT SIDE!

The French ecologist André Gorz says unemployment in the rich countries suggests that less time is now required to produce the *necessities* of life.

IF EVERYONE WORKED LESS WE COULD HAVE FULL EMPLOYMENT

RIGHT TO WORK

Gorz argues that we can live better by working and consuming less – as long as we produce good quality, durable things that are not harmful and do not create resource scarcity when generally available. In fact, the only things *worth* producing are those which remain good for everyone when everyone has access to them . . . things which do not create privilege for some at the expense of others.

FRESH FOOD?

TRAINS? POWER TOOLS? HI-FI?

FAIR-PRICED HOUSING?

BIKES? TOUGH SHOES?

POCK CALCUL?

To escape the eco-crisis and beat technofascism, says Gorz, we must *redirect* production instead of increasing it, making other things in other ways.

A SMALL SELECTION OF BASIC GOODS

DESIGNED FOR MAXIMAL SATISFACTION

AND MAXIMAL ECO-CARE

INSTEAD OF MAXIMAL COMPANY PROFIT

Wage-labour production would meet society's basic needs while 'informal' production – people doing work for themselves, swapping with friends and neighbours, etc – would provide a vast array of goods and services over and above the necessities.

Also, says Gorz, it would help us to move away from the division of labour and allow us to choose different levels of consumption and different lifestyles.

INSTEAD OF SOMEONE CHOOSING THEM FOR US

DIFFERENT STROKES FOR DIFFERENT FOLKS

Near & Far

Decentralisation makes it easier for people to produce what they consume and consume what they produce. Local production makes sense for a lot of things.

Insofar as the natural resources exist locally transport is saved and therefore energy as well. Food is a good example. Instead of freighting bread, grain, meat and vegetables up and down the country or importing them . . .

138

Big & Small

But local production isn't always possible – or desirable. Certain raw materials are only to be found in certain areas. And a lot of machinery & parts demand highly specialised and expensive equipment.

SMALL IS BEAUTIFUL

BUT BIG HAS ITS POINTS

Large-scale *can* provide better goods or work safety. Some things cannot be or should not be decentralised or scaled down.

MINES AND QUARRIES?

IRON AND STEEL, PAPER, CEMENT PROCESSING?

FOREST, RNERS?

THINK GLOBALLY

BUT ACT LOCALLY

Resources must be shared, locally, regionally, nationally and internationally, producing things that others need as well as for ourselves.

A human environment

Norway's political ecologists distinguish between *complex systems* and *complicated systems* when seeking to outline suitable living environments for human beings.

WHERE COMPLEX MEANS DIVERSE

AND COMPLICATED MEANS, HARD TO DEAL WITH

Complex systems
Little dependence on outside world
Self-defined activity
Integrated work
Overall view
Co-determination, responsibility
Symmetrical relationships (mutual exchange)
Cooperation between equals
Local diversity
Majority working in small units
Limited transport needs

THIS MODEL ENCOURAGES SELF-RELIANCE

Complicated systems

Great dependence on outside world
Most activity defined by others
Division of labour
Confusion due to fragmentary, misleading information
Alienation
One-sided relationships (adjustment, exploitation)
Hierarchical relationships, competition
Specialisation
Large units dominate
Extensive transport needs

Ecological planning

Davis, a Californian town of 35,000 inhabitants, was the first in the US to adopt an Energy Conservation Building Code. This led to some ecologically sound planning which radically changed a part of the town.

A whole neighbourhood of 700 people in 250 homes has set out to improve living standards without damaging the environment. Here, self-sufficiency is a keyword and self-involvement means regular meetings to discuss developments.

Almost everything planted in the Davis project is edible — fruit trees, berry bushes, etc. Things are ripening all the year round. There is also a communal compost heap.

Roads are narrow (so small, low-energy cars are preferred) and do not run through living areas. In Davis as a whole, public transport is now well developed.

Eco-Villages are increasingly viewed as a promising solution in town planning, whether on their own in the countryside or clustered together as an Eco-City.

Such communities have been built, are being built or are planned in many countries around the world.

What exactly is an Eco-Village? Robert Gilman* defines it as a
• human-scale
• full-featured settlement
• in which human activities are harmlessly integrated into the natural world
• in a way that is supportive of healthy human development and can be successfully continued into the indefinite future.

The optimal size, he says, is one that allows people to know each other and feel they can influence things. The upper limit would probably be 500 inhabitants. A full-featured settlement means one where all the major functions of normal living — homes, food provision, manufacture, leisure, social life and commerce — are plainly present and in balanced proportions. Thus the eco-village is a microcosm of society as a whole.

* Writing in the US quarterly "In Context", of which he is founding editor. The illustration by Steve Kennel is from the same magazine.

Ecological farming

Modern 'industrial' farming, with its monoculture and biocides, energy-guzzling machinery and fertiliser, cannot feed us in the long run. There are other ways that can . . .

TO GOVERNMENT agronomists, organic farming is a scandalous state of affairs, an incredible step backwards. Think of it: "Farming with manure! It's like going back to the time of Louis XIV!"

In the face of such criticism, Philippe Desbrosses, vice-president of the Federation Nationale d'Agriculture Biologique (FNAB), one of France's biggest organic farming groups, simply gives a Gallic shrug. His farm, near Romorantin in the Loir-et-Cher departement, has been officially recognised as best in the region.

VIVE LA BIOLOGIE!

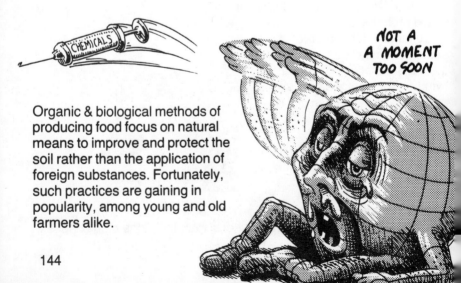

NOT A A MOMENT TOO SOON

Organic & biological methods of producing food focus on natural means to improve and protect the soil rather than the application of foreign substances. Fortunately, such practices are gaining in popularity, among young and old farmers alike.

Countless farmers around the world are switching back to organic methods — or choosing them from the outset — as the price of abusing the soil becomes apparent. Many are able to sell their produce for the same price as the chemical farmer, others who cannot are finding enough consumers willing to pay the extra.

OUR METHODS OFTEN REQUIRE MORE PEOPLE

NICE CHANGE FROM SITTING ALONE ALL DAY

The switch from chemical to organic or biological farming isn't easy, especially if the farm has long been without livestock.

Many farmers experience two or three difficult years before the soil can be restored to natural health.

BUT THEN THINGS QUICKLY GET GOING

MIGHTY MICROBE

PESTICIDES

HERBICIDES

ANTIBIOTIC

In contrast to the Agribusinessman the ecologically-minded farmer is a first-rate technician. He or she masters the demanding technique of creating new and sustainable ecosystems through which life-energy can flow freely and naturally. This requires a thorough understanding of how nature works. For example, you need to know . . .

An alternative farming policy would stress not only sound soil practices but also self-reliance and a radically different overall structure.

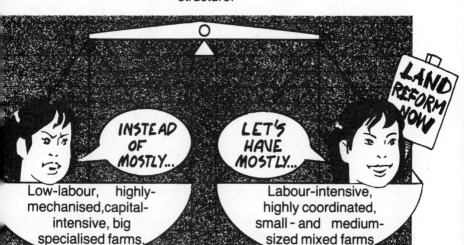

Cleaner & Safer

The sooner we can persuade governments to switch from fossil fuels and nuclear power to energy from the sun, wind, water and biomass, the quicker we can stop pollution and ease the threat to our health, safety and surroundings.

Dismantling military arsenals — the worst single pollutants on Earth — releases funds for the giant task of removing the toxic elements now poisoning the biosphere and building up sustainable energy alternatives.

Recycling

One way to save energy is to take a leaf out of nature's book and keep everything going round and round.

WASTE

COMPOSTING

FERTILISER

WASTE

WASTE

EATING

FOOD

GROWING

COOKING

FOOD

HERE'S A BIG ENERGY SAVER

WASTE

STOP ONE-WAY PACKAGING

Paper, scrap metals, glass, etc, are now widely recycled. Used plastic is being remoulded into planks.

Eco-Technology

Does the future really belong to Flash Gordon? Many scientists and technicians around the world, amateur and professional, are pursuing a path *away* from technofascism. Appropriate . . . soft . . . intermediate . . . alternative . . . convivial . . . radical . . . whatever their technologies are called they share some common features.

WE'RE NOT AGAINST TECHNOLOGY - WE'RE FASCINATED BY IT

ECOLOGICALLY ACCEPTABLE TECHNIQUES

COMBINING OLD AND NEW

ENCOURAGING LOCAL USE AND DEMOCRACY

INSTEAD OF CENTRALISM AND TECHNOCRACY

'Advanced' technology usually means supertankers, missiles, breeders, etc. But what could be more advanced than using technology not just to spare labour but to decentralise and equip society on a human scale for human needs? It's a job with a future . . .

WE CAN NOW DESIGN THE MACHINERY R ELIMINATING SLAVER WITHOUT ENSLAVIN MAN TO THE MACHINE

We have vacancies for green technologists in the fields of transport, work safety, energy production, waste management, fibre research, food production and distribution, environmental control, housebuilding, agriculture, forestry, textiles, electronics, metallurgy, furniture-making, machine design, cybernetics and just about any other field you care to name except mass destruction!

THEY COULD MAKE VIRTUALLY INDESTRUCTIBLE TEXTILES

ECOLOGY AND TECHNOLOGY

AND EASILY REPAIRED MACHINES GOOD FOR 50 OR 100 YEARS

NOT TECHNOLOGY ALONE

An alternative society in solidarity with the Third World would seek to help the poorer countries with things the people there need most — cheap, labour-saving and fuel-saving technology appropriate to village conditions and small-scale cooperative industries, carefully-researched and field-tested.

LIKE GOOD STOVES, GRINDERS, PUMPS AND LOOMS

BETTER LATHES AND PRESSES

The 'ecological perspective' implies global solidarity — that we're all responsible for everyone alive today, for future generations, and for the Earth, which is our common home.

It implies a drastic restructuring of production and consumption patterns and goals to ensure a decent life for the large section of humanity whose basic needs are not being met.

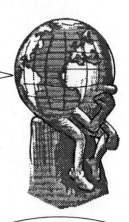

It implies a different set of values — that we must stop measuring people by their 'efficiency' and 'productivity' and start talking about health, harmony, beauty, justice and equality.

Wherever we live, we must decide what is a 'reasonable' living standard and move towards 'sustainable development' as quickly as possible.

Action!

A lot of today's environmental problems are large-scale, continental or global, and require international attention at the highest levels. On the other hand, most of them stem from our lifestyles, political choices and consumer preferences in the industrialised world — so what we do personally, at home, in our local community, at our school or workplace is crucial. In other words, ultimately it's up to you, me and the folks down the road!

Ecology begins at home

To reduce our rubbish (and not squander resources), let's
— use glass returnables instead of aluminium cans and plastic bottles,
— recycle newspapers and magazines,
— avoid paper plates, cups, nappies, towels, etc, aluminium foil and plastic wrapping,
— avoid products with unnecessary packaging,
— compost what we can — in the garden, the cellar or on the balcony*.

We consume more food than anything else. Whenever we buy food we are choosing which kind of farming we want to support now and in the future. Behind almost every item we buy there is a factory, resources, energy and transportation.
Processed foods waste a lot of energy on cooking, cooling, drying, transport, etc, before being re-heated again at home. Better to buy fresh food. Let's also try to
— choose more foods produced closer to home,
— choose more foods that have been naturally grown,
— eat less red meat and broiler chicken (for both health and environmental reasons).

To bring down energy wastage, global inequality and pollution, let's
— go by bike, bus or rail (or on foot) instead of by car,
— look for simpler, safer ways to keep our home, car and clothes clean,
— choose clothing, shoes, tools, toys, furniture, etc, that last and can be repaired,
— finally stop smoking (for the sake of our lungs and the Third World's soil)!

The opportunities are endless.

ADD YOUR OWN SUGGESTIONS HERE

* Books at your library or bookshop can tell you how to make a home compost, even if you only have a space under your sink!

Archie's steps

Moving your life in a different direction, adapting to ecological cycles, takes courage and time. Don't rush it! Take it in easy stages, going on to the next one when you can find the courage and resources to do so.

ORGANICALLY GROWN

WHENEVER I CAN FIND THEM

The suggestions on the previous page come mainly from Archie Duncanson, an American living in Sweden who simply set out to eliminate his share of pollution in the world, step by step, and wrote a fascinating little book about it.*

He started with his garbage and ended up changing his life!

You, too, can set your own goals and go at your own pace.

FLOUR-WHOLE WHEAT

BIODYNAMIC DEMETER ASSOC.

EARTH   AIR   SUN   WATER   NOTHING

SPROUTS

SOAK OVERNIGHT, THEN RINSE TWICE DAILY UNTIL DONE

RED CURRANTS ON MY BALCONY

POTATOES FROM THE "ALTERNATIVE" FARMER

EGGS FROM OUTDOOR CHICKENS

*Ecology Begins at Home, obtainable from A. Duncanson, Örnstigen 9, Täby, Sweden.

Joining forces

Joining with others, though, is often easier. GAP is short for Global Action Plan, an international campaign that encourages and helps people all over the world to change their lifestyles, the main goal being to achieve the targets set in the Brundtland Report, *Our Common Future*.

Its central feature is an international reporting and feedback system. Each group of people (eco team) working with the campaign material sends in monthly progress reports describing concrete changes achieved and receives in return a newsletter showing aggregates. Set up an eco team with your friends, neighbours, schoolmates, pupils, colleagues at work!

Write for full details of the Eco Team Workbook to GAP UK, PO Box 893, London E5 9RU, England.

Eco-Politics

Changing our lifestyles is one part of the solution. Bringing pressure to bear on government is another. In the same way as we have to integrate ourselves and the things we do with the ecological cycles of the Earth, ecological thinking has to be integrated into politics. All policy-making must take the environment into account. So...

We're all citizens and consumers...

*"We have
now become a
threat to our own existence.
In this situation, it is just as
dangerous to believe that 'the
experts will sort it out' as to shrug
and say 'we've had it, anyway'.
It's not a case of 'evil people' wanting
to destroy the natural environment
and 'good people' defending it. Good
and evil live side by side in us all.
What each of us needs above all else
is a worthier, more meaningful
concept of reality. One that embraces
all life on Earth, the Earth that we are
all a part of and dependent on.
We didn't inherit the Earth from our
parents. We have borrowed it from
our children.
We have **all** borrowed it and we are
all going to hand it back."*

(The Natural Step)

Eco-demands

- Let basic human needs and ecological care shape production, consumption and lifestyles!

- Meaningful work, a reasonable living standard and a clean environment for all!

- Conservation of energy and raw materials, and sustainable development policies!

- Green laws, levies and taxes to move things in the right direction!

A
Good
Read...

YIPPEE!

BILLY THE BOOKWORM SAYS :–

Books . . .

There are hundreds of books on environmental issues. Among the more interesting and/or inspiring ones are:

The Gaia Atlas of Planet Management
Simply one of the best reference books to sum up the crises facing us and ways out of them.

Working Greener
Kathleen Ralston and Chris Church (GreenPrint Books)
A straightforward 'how to' guide to changing your workplace environment.

Shopping for a better world
(Kogan Page)
One of several books produced by New Consumer magazine, this one describes itself as a 'quick and easy guide to socially responsible shopping' and as such goes several steps beyond simple green consumerism.

Caring for the Earth
(Earthscan/WWF)
An update of the 1984 World Conservation Strategy, this readable document spells out detailed strategies for action at a global, national and local level.

Staying Alive
Vandana Shiva (Zed Books)
Vandana Shiva is one of India's foremost environmental campaigners. Her analysis of environmental problems from a radical Third World woman's viewpoint is challenging and important reading.

Natural Capitalism
Paul Hawkins, Amory B. Lovins (Earthscan/Kogan Page)
An introduction to 'natural capitalism', seeking to show the way to future success in business. It describes the enormous business opportunities from the transformations which are already under way.

Culture and Global Change
Tracey Skelton, Tim Allen (Routledge)
This volume explores aspects of culture and development at a time of rapid global change. Contributors debate the importance of culture to development discourse and the Third World, stressing that there must be a qualitative understanding of the complexities and dynamics of everyday lives.

The Third Revolution (Population, Environment and a Sustainable World)
Paul Harrison (Penguin)
Just as crisis spurred the Agricultural and Industrial Revolutions, so it may initiate the third revolution which is needed to achieve sustainable development. Harrison provides a bulletin on the state of the planet and the process of destruction, and a blueprint for the third revolution.

Life on a Modern Planet (A Manifesto for Progress) – Issues in Environmental Politics
Richard North (Manchester University Press)
This text re-evaluates global questions such as feeding the world, energy, pollution and green consumerism. It argues that the fashionable view that 'progress' opposes 'caring for the environment' should be dropped, and that people should be satisfied with a permanent change in their environment.

Publications . . .

Organisations like Friends of the Earth, Greenpeace, the Green Party, the World Wide Fund for Nature, etc, publish journals for their members. There are several other good publications around.

BBC Wildlife Magazine
Lots of news. In many ways, the best environmental magazine in the UK.
www.bbc.co.uk/animalzone/wildmag.shtml

The Ecologist
Has been at the forefront of the struggle for over three decades. Increasingly international and increasingly radical. Lots of in-depth coverage.
www.theecologist.co.uk

New Scientist
Probably the best coverage of scientific and environmental issues on a weekly basis.
www.newscientist.co.uk

Nature
Well-respected international weekly review of nature-related issues in science.
www.nature.com

Environmental Politics on the Internet
There is a wealth of sites on the Internet related to environmental issues and ecological concerns, from government agencies to individuals raising local issues.

www.redpepper.org.uk has a valuable index of links to a wide variety of 'alternative' sites, and www.foe.co.uk, the Friends of the Earth site, can be a useful starting point.

www.earthforums.com is a good example of an online discussion group based around environmental issues, and again has a detailed links page.

www.cep.unt.edu/enviro.html is a more academic debating forum, focused on environmental ethics. Try also www.forumforthefuture.org.uk

Much of the Internet is dominated by American-based material, and ecological issues are no exception. www.earthtimes.org is a good example.

Lastly, often quoted in this book are Worldwatch studies. Their informative and comprehensive site is at: www.worldwatch.org

Taking action...

You can get involved in campaigns or practical actions through any of the following organisations, which are just a selection of the hundreds active in the UK:

Friends of the Earth

26 Underwood Street, London N1 7JQ Tel 0207 490 1555
One of the biggest environmental groups in the UK and around the world, with such campaigns as Tropical Rainforests, Land Use and Transport, Water and Waste, and Air Pollution and Global Warming.
www.foe.co.uk

Greenpeace

Canonbury Villas, London N1 Tel 0207 865 8100
The best known of the groups, Greenpeace made their name with direct action but are now a broad-ranging organisation as international as the big corporations that are their main targets. Their work focuses predominantly on pollution in all its forms.
www.greenpeace.org.uk

Global Action Plan

8 Fulwood Place, London WC1V 6HG Tel 0207 405 5633
Global Action Plan is an international charity which has developed a number of schemes that help households, work places and schools act more sustainably.   www.globalactionplan.org.uk

Survival International

11–15 Emerald Street, London WC1N 3QL Tel 0207 242 1441
Not strictly an environmental group, Survival's work with indigenous peoples throughout the world means that they play a vital role by helping those who are most at risk from rainforest loss, etc. They have a network of local groups.   www.survival-international.org

Whale and Dolphin Conservation Society

Alexander House, James Street West, Bath BA1 2BT Tel 01225 334 511
An organisation that works to save marine mammals in an effective and up-beat manner. They have a network of local support groups.
www.wdcs.org

Women's Environmental Network

87 Worship Street, London EC2A 2BA Tel 0207 481 9004
Just what it says it is, WEN focuses on empowering women and campaigning on consumer-related issues, notably on the paper and timber industries. Local activity very much encouraged.
wenuk@gn.apc.org

World Development Movement

25 Beehive Place, London SW9 7QR Tel 0207 737 6215
One of the few development groups in the UK also to work effectively on environmental issues. WDM focuses on trade, Third World debt, aid and justice, and is active locally.
www.wdm.org.uk

World Wide Fund for Nature (WWF)

Panda House, Weyside Park, Godalming, Surrey GU7 1XR
Tel 01483 426444
The biggest of the organisations, with a network of groups mainly engaged in fund-raising. Nationally, their work on environmental education is excellent, and they have a good schools network and youth section.
www.wwf-uk.org

Thanks

The sources of information in this book are too numerous to list. Just about everyone from Montaigne to Marilyn Mehlmann seem to have found their way in, often in stolen chunks and usually uncredited. If anyone sees their own stuff here and feels misused, we're sorry — it's just that you expressed things better than we could! Please take it as a compliment. Or as a case of recycling. And if anyone wants to steal bits of this book for similar ends, feel free. It's all in a good cause.

Stephen Croall/William Rankin
Växjö/Paris, January 1992/2000

Credits

Stephen Croall

William Rankin

Stephen Croall is a British writer and journalist living in Sweden. A former Reuter correspondent in Stockholm and London, he has also worked as a musician, an actor, a shoemaker, a teacher and in day nurseries. He is the author of *Nuclear Power for Beginners*.

Has worked in London for *Oz* and the *Radio Times*; in France for *Actuel* and *Echo des Savanes*; in Sweden for *Etc.* and *Dagens Nyheter*. Born in Edinburgh he is presently living in Paris and working as Graphics Editor at a global newspaper.

Index